INTERNATIONAL CENTRE FOR MECHANICAL SCIENCES

COURSES AND LECTURES - No. 77

KIRK C. VALANIS
UNIVERSITY OF IOWA

IRREVERSIBLE THERMODYNAMICS OF CONTINUOUS MEDIA

INTERNAL VARIABLE THEORY

COURSE HELD AT THE DEPARTMENT
OF MECHANICS OF SOLIDS
JULY 1971

UDINE 1971

SPRINGER-VERLAG WIEN GMBH

Originally published by Springer - Verlag Wien New York in 1972

ISBN 978-3-211-81127-6 ISBN 978-3-7091-2987-6 (eBook)

DOI 10.1007/978-3-7091-2987-6

To my wife Lilian

PREFACE

This volume originated in a course of lectures which were given at the International Institute of Mechanical Sciences, Udine, Italy, during the summer of 1971. It deals with the conceptual formalism of internal variables as a means of extending the scope of continuum thermodynamics, to the extent that phenomena associated with heat conduction and dissipation, that accompanies deformation of inelastic media, may now be considered as a routine aspect of the field.

Of particular importance is our proof of existence of entropy under conditions of irreversibility. This, we believe, gives the field a certain mathematical rigor, which it lacked hitherto. Since the appearance of the proof, voices have been raised to the effect that the entropy function is not unique. We sincerely hope that what is meant is that the entropy function can be determined by integration of the first law, only within an arbitrary funcation of the internal variables. This however, does not imply that the entropy, per se, is not unique.

Also appearing, is the new concept of intrinsic time with respect to which a material is cognizant of its past history. Constitutive theories employing this concept apply to materials where prior deformation has a permanent effect on future mechanical response.

The author is indebted to Professor Piotr Perzyna for many enlightening discussions and to Professor W. Olszak, Rector of the International Institute of Mechanical Sciences, for making the lectures possible.

Udine, July 1971

K. C. Valanis

INTRODUCTION.

Any approach to a thermodynamic theory of a continuous medium undergoing large thermomechanical changes relative to an equilibrium state must come to terms with three fundamental questions :

(i) Does an equation of state exist

(ii) Is temperature a physically meaningful and measurable quantity

(iii) Does entropy exist as a thermodynamic function.

These equations do not arise in thermostatics. The existence of equations of state in terms of macroscopic, measurable state variables is vindicated by experiments. The necessity for the existence of temperature and the sufficiency of the latter in conjunction with other state variables, such as strains, to describe uniquely the thermomechanical behavior of a large class of materials appear to be well documented. Finally the existence of entropy at least for systems with two state variables has been proved rigorously.

When transient thermomechanical conditions prevail, however, the above questions are more difficult to answer, and they have been the subject of a great deal of controversy, in the past. See, for instance Ref. 1. Attitudes to the above questions have divided the field into three schools, i.e., those of (a) Onsager (b) Truesdell-Coleman and (c) Meixner ; we would like to consider ourselves as the fourth. Meixner[(1)] and Perzyna[(2)] discussed these schools and we shall, in the following, add our views to the discussion.

The answer to question (i) is complicated by consideration of the size of the thermodynamic system (material region). For the sake of discussion let us introduce the micron (10^{-4} cm) as our unit of measurement. A system whose mean linear dimension, $(\text{volume})^{1/3}$, is 10^5 microns i.e., one centimeter, is a large system. If one chooses to define the internal energy, say, of such a system with two state variables such as density and temperature, then evidently such a definition is physically meaningful if the density and the temperature are uniform throughout the system. For these conditions to be fulfilled, the stress vector on the surface of the system must be normal to the surface, everywhere, and its magnitude must be the same on all parts of the surface, while at the same time, any heat that enters or leaves the system, must do so vanishingly slowly so that conduction has time to reduce the temperature to a constant value throughout the system. This thermal process is known in classical thermodynamic circles as "reversible" addition of heat.

Such constraints on the suitability or otherwise of the state variables, are removed by the notion of the continuum. This is an idealization of matter, which in a finite material region, is assumed to be piece-wise continuous. This notion allows one to reduce the size of thermodynamic system to infinitesimal proportions.

In physical terms, this means that, whatever the density or temperature variation in the above system one may always choose a sufficiently small sub-system such that the state variables are (to great degree of accuracy) uniform over the volume of the sub-system. We choose a convenient minimum mean linear dimension of an infinitesimal sub-system to be the micron. First because the size of our thermomechanical measuring devices is at least as large as a micron, and second because a smaller dimension might tend to make "average" quantities such as heat flux, temperature, and deformation gradient less meaningful. Such a system might be called a microsystem, but we have not adhered too strictly to this nomenclature in our lectures.

There are other advantages which may be reaped from the continuous notion of matter and the resulting microsystem. The conservation laws which are in an integral (global) form when applied to a large system, reduce to a differential (local) form when applied to a microsystem. The kinetic energy term which appears in the conservation of energy equation (First Law of Thermodynamics) for large systems, drops out (as a consequence of Newton's law of motion) when the system is reduced to a microsystem.

As a consequence the need for a "quasistatic" process, when this is introduced for the mere purpose of eliminating the kinetic energy term, no longer exists when the thermodynamic system is a microsystem.

As a result of the above discussion we have reduced question (i) to the following question : "Does an equation of state for a microsystem exist"?

Before we answer this question we make certain observations. From a continuum phenomenological point of view the microsystem is a black box. We can make measurements of its state variables only on its surface, and its equation of state can be disturbed only by means of inputs (such as displacements) on its surface. In some instances the thermodynamic state of the microsystem (i.e., the values of its state functions such as internal energy and stress) is completely defined by the surface values of the state variables (such as displacements), which in turn define equivalent internal state variables (such as the strains).

But this is true only in cases of non-dissipative materials such as elastic materials, inviscid compressible liquids, gases, etc., or in viscoelastic materials where the surface variables change their values at a vanishingly small rate. It is certainly not true in the case of dissipative materials and/or in the presence of rapid rates of change of the surface variables.

In this strict sense one can say that an equation of state for the microsystem does not always exist, in so far as the surface values of the state functions at time t are not uniquely determined by the surface values of the state variables at time t.

Of course a more fundamental model of matter is the atom. Depending on the degree of "fineness" required the microsystem consists of a large number of point masses (atoms, molecules, clusters thereof etc.). The underlined statement above means that the internal energy, say, of the microsystem is not defined by the position of the point masses at the surface of the system but the positions of interior point masses must also be considered.

The displacements of the interior point masses relative to their equilibrium position we shall call internal variables. The characteristic of these variables is that they are not measurable by means of phenomenological experiments ; i.e., they are not observable.

In terms of these as well as the observable variables an equation of state for the microsystem exists.

In the Onsagerist theories(*), at least as interpreted by Meixner(3), Biot(4) and others(5,6), one finds that the internal variables are recognized as necessary additional thermodynamic variables, that must be introduced in the equation of state to describe irreversible effects. However, one sees in addition, that the terminology of "forces" and "fluxes" is introduced together with the essential assumption (if any progress is to be made) that these are linearly related.

We have avoided such relation and have replaced it instead by an "internal constitutive equation", which relates the gradient of the free energy with respect to the internal variables, to the rates of change of the latter. The physical motivation for this approach is given in Section 7.

In the Coleman (8) approach the concept of a state function is abandoned. A distinction is drawn between dependent and independent variables and the former are considered functionals of the real time histories of the latter. The Truesdell-Toupin(9) form of the Clausius-Duhem inequality (**) is utilized to derive constitutive equations in a very general functional form. Because the analytic form of the functionals involved is not known a priori these are approximated by their first Frêchet derivative, an approximation which is predicated on the notion of "fading memory", a concept which was introduced much earlier by Volterra(11), who discussed the same physical problem in purely mechanical terms. One may wish to make the point, at this juncture, that some functionals, simple in form, cannot necessarily be represented with accuracy, by their first Frêchet derivative.

Meixner, quite recently(1), has expressed a disenchantment with internal variables and tried to develop a theory without them. This is essentially a theory of succesive equilibrium states for which one can indeed develop a ther-

(*) In this connection see Kluitenberg, Ref. 7.
(**) Following Eckart (10).

modynamic theory without the use of internal variables ; it applies to situations where current rates of change are slow and the thermodynamic state is near equilibrium. Indeed (for a wide class of materials), when a system is in equilibrium, the observable state variables are sufficient to define its state functions, and the "path" in the observable state space is not necessary. Such a theory, however, is per force limited in its applicability to such special situations.

At this point we turn our discussion to the question of temperature, i.e., question (ii) at the beginning of the Introduction. This question is really in two parts, (a) is temperature physically meaningful in the presence of irreversibility and (b) how accurate is its measurement if (a) is true. Meixner's negative comments in Ref. 1, seem to involve the measurability of temperature rather than its existence. In other words his arguments do not negate the existence but rather the measurability of temperature. However this is true of practically every quantity in physics. The probe which we use to measure a quantity, will inevitably disturb the conditions under which the quantity exists and hence will change its value during the process of measurement.

Thus, Meixner's[1] arguments involve the accuracy of measurement of temperature under transient conditions. No doubt, the accuracy is much better when static (equilibrium) conditions prevail. However, more sensitive and less bulky thermometers (such as thermocouples) may be developed which will give a more accurate reading of temperature than mercury in glass, or gas, thermometers, used in equilibrium situations.

While convincing arguments regarding the non-existence of the temperature of a microsystem have not been forwarded, in Section 3 we show from the axiom of inaccessibility of thermodynamic states a function which can be identified with temperature, exists.

We finally proceed with question (iii). Arguments regarding the existence of entropy arise, as a result of the restrictive conditions under which, in the past, entropy has been inferred, or shown to exist. These are "quasistatic" conditions pertaining to "reversible processes" and form a very narrow subclass of total class of physical phenomena. As a result the existence of entropy for phenomena which lie outside this sub-class is quite justifiably, open to doubt.

In what we consider as a significant step toward clarifying the situation we prove in Section 3 that a state function which may be identified as the entropy of a microsystem, exists under conditions of irreversibility which can be accounted for, by the use of internal variables.

This is accomplished on the permise of a thermodynamic conjecture, which is a generalized version of a conjecture first given by Caratheodory (12), who gave rigorous grounds for the existence of entropy in the case of systems undergoing a quasistatic process which constitutes a small deviation from an equilibrium state. This he did, by proposing what I will call the Caratheodory Conjecture which may be stated as follows :

> In the vicinity of an equilibrium state of a system there exist other states which cannot be reached quasistatically by processes which are reversible and adiabatic.

The applicability of this conjecture is limited to quasistatic processes that can be regarded as successions of equilibrium states. This allows one to introduce an equation of state in terms of the observable variables and temperature.

However we now realize that the thermodynamic state of an irreversible system can be defined by means of the internal variables. This opens the way for a conjecture far broader than Caratheodory's which we state as follows :

> In the vicinity of a thermodynamic state of a system there exist other states which are not accessible by processes which are reversible and adiabatic.

It is important to note that in our conjecture, we no longer stipulate that inaccessibility be relative to an equilibrium state, and that it be restricted to a quasistatic process. As a result the consequences of our conjecture are far more general. We are led to believe, therefore, that the internal variable theory of irreversible thermodynamics rests on conceptually sound foundations and can serve as a basis for the development of thermodynamical constitutive equations for dissipative materials. We shall present, with certain additions, generalizations and refinements,our work which has already appeared in Ref.'s 13 - 18.

SECTION 1.

Primitive Notions. First Law of Thermodynamics.

One begins by applying the theory of thermodynamics to a "system". In the present case, the system is a material region R_χ the position and configuration of which are defined (*) with respect to an inertial cartesian frame χ^a with a Euclidean metric $\delta_{\alpha\beta}$. This frame will be referred to as the "material frame".

The material region will be regarded as a continuum in the sense that a vanishingly small volume ΔV of the region has a mass Δm, where

$$\Delta m = \rho_o \ (\chi^a) \ \Delta V \tag{1.1}$$

such that the "density", $\rho_0(\chi^a)$, is a piecewise continuous function of χ^a, in the region R_χ.

The region R_χ as well as all its subregions, i.e., the system and its subsystems will be considered closed in the sense that they cannot exchange matter, through their boundary, with their surroundings. In the light of this hypothesis, the principle of conservation of mass is trivially satisfied. A local subsystem will be one that is bounded by an orthogonal infinitesimal parallelpiped.

It will be taken as an axiom that the system possesses internal energy E. A purely thermal process is one, during which energy flows through the boundary of the system and/or is supplied to its interior in the absence of body forces, when part of the boundary is contrained in a fixed position, and there are no surface forces acting on the remainder of the boundary. This type of energy will be called heat and will be denoted by Q. A process during which heat is supplied

---

(*) At some convenient time, say $t = o$. This configuration will be referred to as the reference configuration,

to the system by flow through its boundary, is called conduction. In the absence of conduction, energy supplied in the interior of the system during a thermal process will be called heat supply. An adiabatic boundary is one through which no heat can flow. A boundary that allows heat flow is diathermal.

An adiabatic process (*) is one during which, energy (in the absence of heat supply) is transmitted to the system with an adiabatic boundary, by the application of surface or body forces. This type of energy will be called work and will be denoted by W.

A process during which work is done on the system while heat is being transmitted to the system will be called a thermomechanical process.

The energy present in a system by virtue of its motion is called "kinetic energy" and will be denoted by K.

First Law of Thermodynamics. This law is a statement to the effect that the increment of ΔW of work done on the system plus the increment ΔQ of heat received by the system is equal to the increment ΔK of kinetic energy plus the increment ΔE of internal energy of the system i.e.,

$$\Delta E + \Delta K = \Delta W + \Delta Q \tag{1.2}$$

irrespective of the magnitude of the increments. Thus this law is an assertion of "conservation of energy". If the increments are infinitesimal, eq. (1.2) may be written in the "rate" form.

$$\dot{E} + \dot{K} = \dot{W} + \dot{Q} \tag{1.3}$$

where a dot over a quantity denotes its derivative with respect to time, assuming of course that quantities in eq. (1.3) are continuous functions of time.

(*) This process will also be called purely mechanical.

The internal energy is an <u>extensive</u> quantity, in the sense that there exists an integral energy density $\epsilon(\chi^a)$ (per unit mass) piecewise continuous in χ^a, time that

$$E = \int_{V_0} \rho_0 \epsilon dV \tag{1.4}$$

where V_0 is the volume of the region in the reference configuration. The kinetic energy is also an extensive quantity.

<u>Application of the Fist Law to a Continuous Medium.</u> It is observed experimentally that when a material region (system) R_χ, whose position and configuration at time $t = o$ is defined with respect to the cartesian coordinate system χ^a is acted upon by surface forces and/or long range forces and/or if it is supplied with heat through a diathermal boundary or by radiation, it will, in general, acquire motion.

This motion is conveniently described in a cartesian coordinate frame (*) y_i – known as the spatial system – through the set of relations

$$y_i = y_i(\chi^a, t) \tag{1.5}$$

For the sake of subsequent discussion a point with coordinates χ^a will be referred to as a "particle" whereas a point with coordinates y_i will be referred to simply as a "point".

The velocity v_i of a particle is defined as

$$v_i = \left.\frac{\partial y_i}{\partial t}\right|_{\chi_a} \equiv \frac{Dy_i}{Dt} \tag{1.6}$$

(*) With a Euclidian metric δ_{ij}. This frame may be regarded as the laboratory frame in which the motion is observed. It is then the observer's frame of reference and is completely arbitrary as far as its position and orientation are concerned with respect to the material frame χ^a.

The operation $\frac{D}{Dt}$ is known as the "material derivative". The acceleration a_i, of a particle is defined as

$$a_i = \frac{Dv_i}{Dt} \tag{1.7}$$

The kinetic energy K of the system is defined as

$$K = \frac{1}{2} \int_V \rho v_i v_i dV \tag{1.8}$$

When the heat flow is spatially non-uniform, it may be expressed most conveniently in terms of the heat flux vector h_i. The direction and magnitude of the vector denote, respectively, the direction of flow and the quantity of heat flows, per unit time, through a unit area normal to the direction of flow.

Let T_i denote the stress field which acts on the surface S of the region R_χ and f_i the body force field (per unit mass) acting in the interior of R_χ. Also let q be the heat supplied to the interior of the body per unit mass. Then in terms of the previously defined quantities, the First Law may be written as follows for the system R_χ :

$$\begin{aligned} \frac{D}{Dt} \int_V \rho \epsilon dV + \frac{D}{Dt} \int_V \frac{1}{2} \rho v_i v_i dV &= \int_S T_i v_i dS + \int_V \rho f_i v_i dV - \\ &- \int_S h_i n_i dS + \frac{D}{Dt} \int_V \rho q dv \end{aligned} \tag{1.9}$$

where V and S qre, respectively, the mappings of the material volume and the material surface of R_χ, in the spatial frame y_i. This form of the First Law will be referred to as the "global form". It is shown in text-books on continuum mechanics that eq. (1.9) in conjunction with the Laws of conservation mass, linear momentum and angular momentum reduces to the following local form :

$$\rho \frac{D\epsilon}{Dt} = T_{ij} v_{j,i} - h_{i,i} + \rho \frac{Dq}{Dt} \tag{1.10}$$

where a comma following a subscript denotes differentiation with respect to the corresponding spatial coordinate and T_{ij} is a symmetric tensor. Equation (1.10) is the necessary and sufficient condition that the energy of a local subsystem be conserved.

It is important to note that eq. (1.10) is valid whether the process is "quasistatic", dynamic, "reversible" or "irreversible". This must be emphasized because eq. (1.10) will play a central role in the proof of existence of entropy and subsequently, temperature.

Equation (1.10) may also be written in the material system χ^a, as

$$\dot{\epsilon} = \tfrac{1}{2} \frac{\rho_o}{\rho} \tau^{a\beta} \dot{C}_{a\beta} - h^a{}_{,a} + \dot{q} \tag{1.11}$$

where, now, ϵ and q are calculated per unit undeformed volume $\tau^{a\beta}$ is the stress tensor in the material system, i.e.,

$$\tau^{a\beta} = \chi^a{}_{,i} \chi^\beta{}_{,j} T_{ij} \tag{1.12}$$

and h^a is the heat flux vector in the material system, calculated per unit undeformed area, i.e.,

$$h^a = \frac{\rho}{\rho_o} \chi^a{}_{,i} h_i ,$$

To simplify the discussion in the subsequent Sections a vector notation has been adopted, where

$$\frac{\rho_0}{\rho}\left(\tau_{11}\,,\ \frac{\tau_{12}}{\sqrt{1/2}},\ \frac{\tau_{13}}{\sqrt{1/2}},\ \tau_{22}\,,\ \frac{\tau_{23}}{\sqrt{1/2}},\ \tau_{33}\right) \equiv (\chi_1\,,\ \chi_2\,,\ \chi_3\,,\ \chi_4\,,\ \chi_5\,,\ \chi_6) \tag{1.13}$$

$$\tfrac{1}{2}\left(C_{11}\,,\ \frac{C_{12}}{\sqrt{1/2}},\ \frac{C_{13}}{\sqrt{1/2}},\ C_{22}\,,\ \frac{C_{23}}{\sqrt{1/2}},\ C_{33}\right) \equiv (x_1\,,\ x_2\,,\ x_3\,,\ x_4\,,\ x_5\,,\ x_6) \tag{1.14}$$

Also

$$-\,h^{a}{}_{,a} + \dot{q} \equiv \dot{Q} \tag{1.15}$$

where Q is the heat supplied to a unit undeformed volume of material per unit time. This notation may conflict somewhat with that of eq. (1.2) where Q was the heat supplied to a large region, but no confusion is likely to arise.

In this notation eq. (1.11) becomes :

$$\dot{\epsilon} = \chi_i \dot{x}_i + \dot{Q} \tag{1.16}$$

Note that an adiabatic process implies that $\dot{Q} = 0$.

This is the first law of thermodynamics for an infinitesimal material system undergoing a general thermomechanical process, not necessarily quasi-static or reversible. Equation (1.16) will play a central role in the proof of existence of entropy.

SECTION 2.

State Variables and Thermodynamic State ; Entropy of a Gas.

From now on a local subsystem will be referred to as a "system" unless otherwise stated.

Definition : A state variable of a thermodynamic system is a measurable (*) physical entity which may expressed as a continuous function of time and is such that its numerical value at time t is necessary (but not always sufficient) to define uniquely the numerical value of the integral energy density ϵ at time t.

Under these conditions ϵ is said to be a function of the state variable.

As a direct consequence ot the above definition, if changing the numerical value of a physical entity brings about a change in the value of the internal energy density, then such a entity is a state variable. If, furthermore, this state variable is not a function of other state variables, previously found, it is said to be "primitive", relative to the other variables.

By acknowledging the feasibility of changing the internal energy of a system by a purely mechanical process on finds that the strain components $C_{\alpha\beta}$ are primitive state variables. Also a recognition that ϵ may be changed (by means of a thermal process) while $C_{\alpha\beta}$ remain constant shows that $C_{\alpha\beta}$ are not sufficient to uniquely determine ϵ. Another primitive state variable is necessary which is associated with the degree of "coldness" or "hotness" of the system. This added variable is, of course, the temperature T of the system.

(*) Directly or indirectly

One may perform experiments which show that the stress components $\tau^{\alpha\beta}$ are also functions of $C_{\alpha\beta}$ and T.

A set of primitive state variables, the numerical values of which at time t are necessary and sufficient to determine uniquely the numerical values of ϵ and $\tau^{\alpha\beta}$ at the time t, is said to be a "complete set". A complete set of primitive state variables defines uniquely the "thermodynamic state of the system" ; ϵ and $\tau^{\alpha\beta}$ are then said to be "state functions" of the set.

Nonetheless, one must bear in mind that under certain "thermodynamically restrictive" circumstances (such as adiabatic conditions) a primitive variable such as temperature may become a function of other primitive variables, such as the strain components $C_{\alpha\beta}$.

Reversible systems.

By definition, a system is said to be reversible if its internal energy density ϵ, and stress components $\tau^{\alpha\beta}$ are state functions of the components of strain $C_{\alpha\beta}$ and the temperature T.

Irreversible systems.

By definition, a system is said to be irreversible if the current value of the internal energy density as well as the current values of the components of the tensor $\tau^{\alpha\beta}$ cannot be determined uniquely from the current values of the strain components $C_{\alpha\beta}$ and the temperature T ; evidently, for such systems ϵ and $\tau^{\alpha\beta}$ are not state functions of $C_{\alpha\beta}$ and T.

Reversible Process.

A reversible system is said to undergo a reversible process if its thermodynamic state is being changed under adiabatic conditions.

One of the main tasks of thermodynamics is to establish which quantities are state functions and which are not and to establish useful relationships between those that are.

To fix ideas we consider the case of a perfect gas. By definition, such a system has the following properties.

(i) It can withstand only uniform hydrostatic pressure, i.e.,

$$T_{ij} = - p\delta_{ij} \tag{2.1}$$

where p is positive.

(ii) The internal energy density is a state function of temperature only ; in particular

$$\epsilon = C_v T \tag{2.2}$$

where C_v is a constant, known as the specific heat at constant volume.

(iii) The pressure of the gas is a state function of its volume and temperature through the "constitutive equation"

$$pV = RT \tag{2.3}$$

where R is a constant which will depend on the reference configuration. Evidently p is a state function of V and T. Therefore, the perfect gas is a reversible system.

For the sake of simplicity let us assume that heat is supplied to the gas by heat sources and heat sinks inside the gas. The first Law for this system becomes ;

$$C_V dT + p dV = dQ \tag{2.4}$$

Equation (2.4) needs careful discussion. The differential quantity dQ is the numerical value of the heat supplied to the system during the infinitesimal time interval dt, i.e.,

$$dQ = Q(t + dt) - Q(t). \tag{2.5}$$

Because Q is assumed to the continuous function of t, one is justified in writing

$$dQ = \dot{Q} dt \tag{2.6}$$

in the sense that given a positive number δ, however small, then a small time interval dt may be found such that

$$\left| dQ - \dot{Q}\, dt \right| < \delta \tag{2.7}$$

The differential quantities dT and dV should be given the same interpretation. Thus equation (2.4) is a numerical relation and is not to be taken as a presumption of any functional relation between Q, T and V.

Consider a thermodynamic process during which the volume and temperature of the system depart from their initial values at time t = o, but they return to these values at the end of the process. This process constitutes a "thermodynamic cycle" at the end of which the state variables assume their initial values.

Such a cycle is shown in Figure 1.

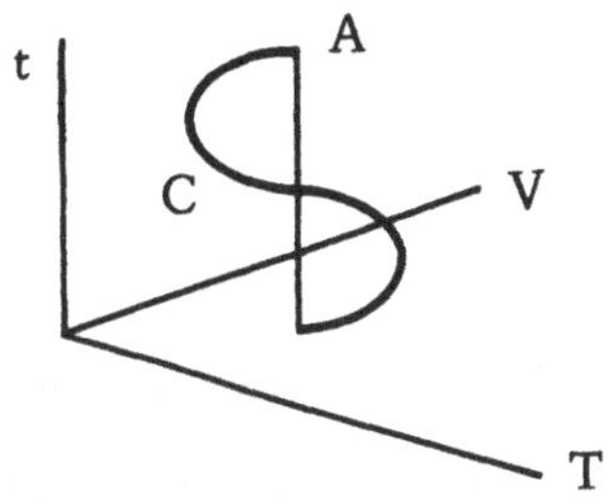

Fig. 1. A thermodynamic Cycle.

As a result of the continuity assumptions on V and T,

$$\int_{o}^{t_a} (C_V\dot{T} + p\dot{V})dt \equiv \oint (C_V dT + pdV) = W \tag{2.8}$$

where $\oint$ means integration following the variation of the state variables V and T on the curve C and W is the work done (during the cycle) which, of course, is not always zero.

Furthermore, as a result of equation (2.4)

$$W = \int_{o}^{t_a} \dot{Q}\, dt = \oint dQ \tag{2.9}$$

Equation (2.9) shows conclusively that Q is not a state function of V and T otherwise $\oint dQ$ would always be zero. We conclude therefore that the heat Q is not a state function of the system.

On the other hand equations (2.3) and (2.4) give,

$$C_V \frac{dT}{T} + R\frac{dV}{V} = \frac{dQ}{T} \tag{2.10}$$

The left hand side of equation (2.10) may be written as a differential of a state function η of T and V where,

$$\eta = \log(T^{C_V} V^R) + \eta_o \tag{2.11}$$

In this sense the left hand side of equation (2.10) is integrable.

Furthermore, as a result of equation (2.10)

$$\frac{dQ}{T} = d\eta \tag{2.12}$$

i.e. $\frac{dQ}{T}$ is the total differential of a state function of V and T.

Now let $\int \frac{dQ}{T}$ be the Stieljes integral representation of the Riemann integral $\int_0^t \frac{Q(t)}{T} dt$; then, as a result of equation (2.12)

$$\int \frac{dQ}{T} = \eta(T, V) \tag{2.13}$$

i.e. $\int \frac{dQ}{T}$ is a state function of T and V and is known as Entropy.

Thus for a system which is a perfect gas, internal energy and entropy are state functions of the state variables V and T whereas the quantity of heat transmitted to (or removed from) the system is not.

The assumption that internal energy and pressure are state functions (of a particular type) of the state variables V and T leads to the integrability of the differential form of the First Law. The integrability leads to the establishment of another state function called entropy.

The question is, to what extent do these conclusions retain their validity for other systems. We shall examine this question in its full generality in Section 3.

SECTION 3.

Pfaffians. Theorem of Caratheodory. Second Law of Thermodynamics and Existence of Entropy.

Let a χ-space be the n-dimensional Euclidian space of n independent variables χ_i. Let C be a curve in this space and let it be given in the parametric form :

$$\chi_i = \chi_i(t) \tag{3.1}$$

where $\chi_i(t)$ are continuous functions of t. An infinitesimal element of length ds of the curve C is given by the expression

$$ds = (\chi_i \chi_i)^{1/2} dt \tag{3.2}$$

The components of a vector, length ds, and tangent to the curve are $d\chi_i$. Evidently

$$d\chi_i = \chi_i(t + dt) - \chi_i(t) \tag{3.3}$$

Let $X_i(\chi_k)$ be n single-valued functions of χ_k, in some finite domain of definition of the variables χ_k and let dQ be the numerical value of the scalar differential form $\sum_i X_i d\chi_i$, i.e.,

$$\sum_{i=1}^{n} X_i d\chi_i = dQ \tag{3.4}$$

Evidently dQ tends to zero, as $d\chi_i$ tends to zero for all i.
The differential form dQ is known as a Pfaffian.

A Pfaffian form is said to be integrable if there exist functions

$\theta(\chi_k)$ and $\eta(\chi_k)$ such that

$$\frac{1}{\theta}\sum_{i=1}^{n} X_i \, d\chi_i$$

is equal to the total differential of the function $\eta(\chi_i)$, i.e.,

(3.5) $$\frac{1}{\theta}\sum_{i=1}^{n} X_i \, d\chi_i = d\eta(\chi_i)$$

In this event,

(3.6) $$X_i = \theta \frac{\partial \eta}{\partial \chi_i}$$

and

(3.7) $$\frac{dQ}{\theta} = d\eta(\chi_i)$$

Furthermore if C is a closed curve in a sub-domain of the χ-space over which X_i are single-valued, then

(3.8) $$\int_C \frac{dQ}{\theta} = 0$$

The function θ is called the integrating denominator of the Pfaffian.

An important property of Pfaffian forms is that they are not always integrable in the sense that they do not always admit an integrating denominator ; a powerful necessary and sufficient condition for their integrability, is given by the "Theorem of Caratheodory" :

Theorem : A necessary and sufficient condition that a Pfaffian differential form be integrable in the sense that

(3.9) $$\sum_{i=1}^{n} X_i \, d\chi_i = \theta(\chi_i) d\eta(\chi_i)$$

is the following : Given a point P in the n-dimensional space of coordinates χ_i there are points P' in the neighborhood of P which cannot be connected to P by means of curves that are solutions of eq. (3.10)

$$\sum_{i=1}^{n} \chi_i d\chi_i = 0 \tag{3.10}$$

The consequences of this theorem are far-reaching.

<u>Pfaffian forms and First Law of Thermodynamics</u> :

The First Law of thermodynamics for an infinitesimal sub-system is given by eq. (1.16) and may be written in the "rate form"

$$\dot{\epsilon} - X_i \dot{x}_i = \dot{Q} \tag{3.11}$$

or in the incremental form

$$d\epsilon - X_i \, dx_i = dQ \tag{3.12}$$

where, $d\epsilon = \epsilon(t + dt) - \epsilon(t)$, etc.

The left hand side of eq. (3.12) is indeed, a Pfaffian form, which represents, thermodynamically, the increment of internal energy density of the system minus the increment of work done on the system ; the numerical value of the Pfaffian is equal to the increment of heat supplied to the system.

In this Section we shall be concerned mainly with the conditions under which the left hand side of eq. (3.12) is integrable ; if it is, and to avoid repetitious statements, we shall say that "the first law is integrable" ; the negation of this statement will be used if the left hand side of eq. (3.12) is not integrable.

In the case of the perfect gas, where the constitutive equations for ϵ

and χ_i were given by eq.'s (2.1), (2.2) and (2.3), it was found that the first law is integrable and the integrating denominator was the temperature of the gas. The integrability of the first law led to the existence of another state function called entropy.

The question that still remains is whether the first law is integrable for more general thermodynamic systems ; if so, then the existence of entropy for such systems will have been shown.

Integrability of the first law. Reversible systems.

We shall address ourselves first to reversible systems. It will be recalled that these systems are such that

(3.13 a,b) $$\epsilon = \epsilon(\chi_i\ , T)\ ;\quad X_i = X_i(\chi_k\ , T).$$

The first law for such systems becomes :

(3.14) $$\left[\frac{\partial \epsilon}{\partial \chi_i} - X_i\right] d\chi_i + \frac{\partial \epsilon}{\partial T}\, dT = dQ$$

According to Caratheodory's theorem, the Pfaffian given by the left hand side of eq. (3.14) is integrable if in the seven-dimensional space of the variables χ_i and T there exists points in the neighborhood of a point P which cannot be connected to P along curves of the solution of the equation

(3.15) $$\left(\frac{\partial \epsilon}{\partial \chi_i} - X_i\right) d\chi_i + \frac{\partial \epsilon}{\partial T}\, dT = 0$$

Eq. (3.15) implies that changes in the system can take place only adiabatically ; because the system is reversible such changes are reversible, by definition. Hence, by our conjecture there exist "adjacent" thermodynamic states

$(\chi_i + d\chi_i, T + dt)$ which are not accessible (*) from the state (χ_i, T). Hence, by Caratheodory's theorem, the Pfaffian on the left hand side of eq. (3.15) is integrable. Furthermore, as a consequence of Addendum II of Section 3, there exists a relation

$$T = T(\chi_i) \qquad (3.16)$$

i.e., the temperature of the system cannot be changed unless one varies χ_i, since diathermal processes are not admissible.

Eq. (3.16) defines a hypersurface in (χ_i, T) space. It follows from Addendum II of Section 3, that such a surface is isentropic.

As a result of the above conclusions we may now state the following theorem :

Theorem : If ϵ and X_i $(i = 1, 2.....6)$ are state functions of the state variables χ_i $(i = 1, 2.....6)$ and T only, the Pfaffian form (3.4) is integrable.

Therefore, there exists an integrating factor $\theta(\chi_i, T)$ such that

$$\left[\frac{\partial \epsilon}{\partial \chi_i} - X_i\right] d\chi_i + \frac{\partial \epsilon}{\partial T} dT = dQ = \theta\, d\eta \qquad (3.17)$$

where η is a state function of χ_i and T. As a result,

$$\frac{\partial \epsilon}{\partial \chi_i} - X_i = \theta \frac{\partial \eta}{\partial \chi_i} \qquad (3.18)$$

$$\frac{\partial \epsilon}{\partial T} = \theta \frac{\partial \eta}{\partial T} \qquad (3.19)$$

The quantity η is known as the entropy of the system and is a state function of χ_i and T (**).

(*) Along curves in χ- space that are solutions of eq. (3.15).

(**) I am not aware of a proof such as this in the literature. Furthermore this proof nowhere stipulates that thermodynamic changes suggested by eq. (3.17) need be "quasistatic" or need represent successions of equilibrium states.

One now carries out a simple "point transformation" of the space into itself, i.e.,

(3.20a) $$\chi_i = \chi_i$$

(3.20b) $$\theta = \theta(T, \chi_i) ;$$

one notes that for fixed χ_i , θ is a function of T, i.e., θ has the significance of an intrinsic measure of temperature ; it is necessary that $\left.\frac{\partial \theta}{\partial T}\right|_{\chi_i} \neq 0$. In view of this result and as a consequence of eq.'s (3.20) one immediately obtains the basic equations relating the state functions to the state variables of the system :

(3.21a) $$\frac{\partial \epsilon}{\partial \chi_i} - \theta \frac{\partial \eta}{\partial \chi_i} = X_i$$

(3.21b) $$\frac{\partial \epsilon}{\partial \theta} = \theta \frac{\partial \eta}{\partial \theta} ,$$

where ϵ and η are now functions of χ_i and θ. Further, if one introduces another state function ψ, the free energy density, such that

(3.22) $$\psi = \epsilon - \eta\theta$$

then in terms of ψ eq.'s 3.21 become

(3.23a) $$X_i = \frac{\partial \psi}{\partial \chi_i}$$

(3.23b) $$\eta = -\frac{\partial \psi}{\partial \theta} .$$

Also, one finds the very important relationship :

(3.24) $$\dot{\psi} = X_i \dot{\chi}_i - \eta \dot{\theta}$$

which implies that under isothermal conditions ($\theta = 0$), ψ is the amount of energy of the system which can be converted into work.

We summarize the above results by means of the following lemma :

Lemma : If the internal energy ϵ of a system and the mechanical "forces" (stresses) X_i acting on the system are functions of the "strains" χ_i (such that $X_i \dot{\chi}_i = \dot{W}$) and the temperature θ, then as a consequence of our conjecture, and Caratheodory's theorem there exists a state function $\eta(\chi_i, \theta)$ called entropy such that

$$d\eta = \frac{dQ}{\theta} \tag{3.25}$$

Definition : Systems for which eq. (3.25) is true are reversible systems.

In some textbooks eq. (3.25) is referred to as the Second Law of Thermodynamics. We shall withhold comment on this point at this time. As will be seen shortly, the integrability of the First Law does not necessarily imply eq. (3.25). Some uncertainty exists on this point and the reader will be advised to read the following discussion very carefully.

Irreversible Systems.

In such systems the internal energy ϵ and the forces X_i are not uniquely determined by the state variables χ_i and T of the system. For instance in some materials (commonly known as viscoelastic) the current state of stress will depend (*) on the whole history of their deformation. Thus the current values of

(*) Except in the case of very slow or very fast motion, when such materials may be regarded approximately as reversible systems.

the state variables χ_i and T are not sufficient to define uniquely the thermodynamic state of the system, and additional variables must be introduced to define the state of the system. That these additional variables are sufficient to define the state of the system is, I believe, the fundamental axiom of irreversible thermodynamics.

Axiom : A complete set of primitive state variables (not all observable) can always be found to define uniquely the thermodynamic state of an irreversible system.

The additional non-observable but independent variables required to define uniquely the state of an irreversible system will be called internal variables and will be denoted by $q_\alpha (\alpha = 1, 2.... p)$. Thus in the case of an irreversible system

(3.26 a,b) $$\epsilon = \epsilon(T, \chi_i, q_\alpha) , \quad X_i = X_i(T, \chi_i, q_\alpha)$$

For such a system, the First Law, of course, still applies, i.e.,

(3.27) $$d\epsilon - X_i d\chi_i = dQ$$

However, to my knowledge, the integrability of the first law has not been established for irreversible systems in the past and thus, for such systems, the existence of entropy as a state function has been open to question.

In the following we shall use our conjecture together with the Theorem of Caratheodory (*) to prove the integrability of the first law and thus establish the existence of entropy as a state function for irreversible systems. It is our position that this is a constructive step toward the establishment of irreversible thermodynamics as a rational branch of physics.

(*) A direct proof, without the aid of this theorem, is given in Addendum II.

Evidently, terms of the form $Q_a dq_a$ do not appear in eq. (3.27), where $Q_a =$ $= Q_a(\chi_i, T, q_a)$; the internal coordinates may be viewed as parameters (which may remain constant during some thermodynamic process (*).) Consider the space of the variables χ_i and T. Let P be a point in this space. If there are points P′in the neighborhood of P which cannot be joined to P by means of lines that are solution curves of the differential form,

$$d\epsilon - X_i\, d\chi_i = 0 \tag{3.28}$$

while q_a remain constant then, by virtue of Caratheodory's theorem, the first law is integrable.

<u>However, this is always the case by our Conjecture</u>.

The same conclusion can be reached, on recognizing that while q_a remain constant the system is reduced to reversible one, in which case all that has been said for reversible systems, applies. This recognition, in fact, provides physical justification for our Conjecture.

Thus we have the following Lemma

Lemma : For an irreversible system, with q_a remaining constant, the differential form of the first law is integrable (**).

The consequences of this Lemma are significant. As a result of eq. (3.26) and the integrability of the left hand side of eq. (3.27),

$$\left.\frac{\partial \epsilon}{\partial T}\right|_{q_a} dT + \left.\frac{\partial \epsilon}{\partial \chi_i}\right|_{q_a} d\chi_i - X_i\, d\chi_i = \theta\, d\eta \Big|_{q_a} \tag{3.29}$$

(*) In fact this is a case of a reversible process.

(**) However, eq. (3.27) is not integrable when q_a are varying. This is shown in Addendum III of this Section.

Note that the first two terms if the left hand side of eq. (3.29) do not constitute the total increment of ϵ and therefore the right hand side of eq. (3.29) is not equal to dQ.

Nonetheless there exists an integrating factor $\theta(T, \chi_i, q_a)$ and a state function $\eta(T, \chi_i, q_a)$, again called entropy, such that

(3.30a) $$X_i = \frac{\partial \epsilon}{\partial \chi_i} - \theta \frac{\partial \eta}{\partial \chi_i}$$

(3.30b) $$\frac{\partial \epsilon}{\partial T} = \theta \frac{\partial \eta}{\partial T}$$

Again if we introduce the transformation

(3.31a) $$\chi_i = \chi_i$$

(3.31b) $$q_a = q_a$$

(3.31c) $$\theta = \theta(T, \chi_i, q_a)$$

eq.'(s) (3.30) become :

(3.32a) $$X_i = \frac{\partial \epsilon}{\partial \chi_i} - \theta \frac{\partial \eta}{\partial \chi_i}$$

(3.32b) $$\frac{\partial \epsilon}{\partial \theta} = \theta \frac{\partial \eta}{\partial \theta}$$

where ϵ, χ_i and η are functions of θ, χ_i and q_a where θ plays the role of an intrinsic temperature as in the case of a reversible system.

Furthermore if one introduces another state function, the free energy $\psi(\theta, \chi_i, q_a)$, such that $\psi = \epsilon - \eta\theta$ then in terms of this function eq.'s (3.32) become

(3.33a) $$X_i = \frac{\partial \psi}{\partial \chi_i}$$

(3.33b) $$\eta = -\frac{\partial \psi}{\partial \theta}$$

and as a result of eq.'s (3.33),

$$\dot{\psi} = X_i \dot{\chi}_i + \frac{\partial \psi}{\partial q_a} \dot{q}_a - \eta \dot{\theta}. \tag{3.34}$$

Thus, indeed, under isothermal conditions and in the presence of the additional constraints $q_a = 0$, ψ has the significance of recoverable energy from the system since then

$$\Delta \psi = \int_{\chi_i^{(1)}}^{\chi_i^{(2)}} X_i d\chi_i \tag{3.35}$$

and $\Delta \psi = 0$, if $\chi_i^{(1)} = \chi_i^{(2)}$, where $\chi_i^{(1)}$ and $\chi_i^{(2)}$ are two strain states.

The above results may be put in the form of the following theorem :

Theorem : For an irreversible system undergoing a thermodynamic process, there exists a state function called "entropy" and, as a consequence, a state function called "free energy". Furthermore the free energy function serves the role of a potential from which the forces (stresses) on the system, as well as its entropy,are derivable.

In the past [12], the first part of the above theorem appeared as an axiom. We believe that the proof of the theorem, given earlier, constitutes a contribution to the theory of irreversible thermodynamics.

Thermodynamic Inequalities – The Clausius-Duhem Inequality

It is clear by definition that work cannot be done on a system without deforming its boundary. Then, the following postulate is self-evident.

Under isothermal conditions, the free energy of a system with stationary boundaries cannot increase.

Because, otherwise, this increase in free energy would be converted into work (*) and thus the system could serve as an inexhaustible supply of energy.

The above postulate is a variation of Kelvin's postulate which may be stated as follows :

Under isothermal conditions work cannot be done on a system without disturbing (deforming) its boundary.

To examine the consequence of this postulate consider eq. (3.34) which under isothermal conditions becomes :

$$\dot{\psi} = X_i \dot{\chi}_i + \frac{\partial \psi}{\partial q_a} \dot{q}_a \tag{3.36}$$

Under conditions of constant χ_i (stationary boundary)(**) and as a consequence of the above postulate

$$\dot{\psi}\Big|_{\chi_i, \theta} \leq 0 \tag{3.37}$$

Hence as a result of eq. (3.33) the following inequality holds :

$$\frac{\partial \psi}{\partial q_a} \dot{q}_a \leq 0 \tag{3.38}$$

Let

$$\frac{\partial \psi}{\partial q_a} \dot{q}_a = -\theta \dot{\gamma} \tag{3.39}$$

where

$$\dot{\gamma} \geq 0. \tag{3.40}$$

(*) Totaly, if q_a=constant

(**) Except possibly for rigid body motion.

The ineq. (3.38) is identically satisfied. As a result of the first Law and eq.' (3.33)

$$dQ = \theta(d\eta - d\gamma) \tag{3.41}$$

or

$$d\eta = \frac{dQ}{\theta} + d\gamma \tag{3.42}$$

Equation (3.42) suggests for γ, the name of "irreversible entropy". Evidently, from (3.40)

$$d\eta \geq \frac{dQ}{\theta} \tag{3.43}$$

This last equation is known as the Clausius-Duhem inequality.

Addendum I :

The case of workless strains. The local subsystem is of course a part of the material region R_χ which may subject to strain field not constant throughout R_χ. One may therefore ask whether strain gradients affect the state of the system. Let us call the strain gradients χ_i^1. Since these are workless, but independent, (*) variables they may be viewed at parameters held constant during an incremental thermodynamic process. Using an argument similar to the one concerning the internal coordinates, one may easily establish the existence of entropy and introduce the free energy density ψ which now may depend on χ_i^1.

Under isothermal conditions and in terms of ψ one is lead to the relations

$$X_i = \frac{\partial \psi}{\partial \chi_i} \tag{3.44}$$

(*) Otherwise if they depended on χ_i, say, they could be eliminated as state variables.

$$\frac{\partial \psi}{\partial \chi_i^1} \dot{\chi}_i^1 = 0 \tag{3.45}$$

Two possibilities exist. Either $\dot{\chi}_i^1$ may be prescribed independently of χ_i^1 in which case

$$\frac{\partial \psi}{\partial \chi_i^1} = 0 \tag{3.46}$$

or there exists a kinematic constraint of the type

$$\dot{\chi}_i^1 = f_i(\chi_i, \chi_i^1) \tag{3.47}$$

However, for $\dot{\chi}_i^1 = 0$, eq. (3.47) implies a relationship between χ_i and χ_i^1 contrary to the original assumption. Thus eq. (3.46) must hold with the implication that ψ is independent of workless strains. Similar arguments may be offered for higher order strain gradients. Furthermore other types of constraint exclude histories which are certainly possible under experimental conditions and as a consequence they will not be considered.

Exactly analogous sityations arise when temperature gradients are included as state variables.

Addendum II :

A direct proof of the integrability of the First Law for reversible systems.

We proceed to establish a necessary and sufficient condition for the integrability of the Pfaffian differential form

$$X_i d\chi_i . \tag{3.44}$$

Theorem : A necessary and sufficient condition that the Pfaffian differential from (3.44) be integrable is that the independent variables χ_i are connected by a functional relation of the form

$$f(\chi_i) = 0 \,. \tag{3.45}$$

Proof of necessity. Let eq. (3.44) be integrable. Then by definition

$$X_i = \theta \frac{\partial \eta}{\partial \chi_i} \,. \tag{3.46}$$

Hence as a result of eq. (3.44)

$$d\eta = 0 \tag{3.47}$$

or

$$\eta(\chi_i) = \text{constant} \tag{3.48}$$

which shows that a condition of the form (3.45) is necessary.

Proof of sufficiency. Let eq. (3.45) be true. Then, in the n-dimensional Euclidian space of the variables χ_i (with metric δ_{ij}), eq. (3.45) defines a hypersurface. Consider now a mapping of this n-dimensional space onto itself, given by the relations

$$X_i = X_i(u^1, u^2, \ldots u^{n-1}, \xi) \tag{3.49}$$

where u^a and ξ are independent variables, or conversely,

$$u^a = u^a(\chi_i)$$

$$a = 1, 2 \ldots (n-1) \tag{3.50}$$

$$\xi = \xi\,(\chi_i) \tag{3.51}$$

Let u^a constitute coordinates that lie in the surface and let ξ be a coordinate normal to this surface in the sense that

$$\frac{\partial u^a}{\partial \chi^i} \cdot \frac{\partial \xi}{\partial \chi^i} = 0 \tag{3.52}$$

and furthermore let the equation of the surface $f(\chi_i) = 0$ be described by the set of n eq.'s

$$\chi_i = \chi_i(u^a, \xi_o) \tag{3.53}$$

where ξ_o is a fixed value of ξ.

The vector X_i may now expressed in the new coordinate system by the relations

$$X_a = X_i \frac{\partial \chi_i}{\partial u^a} \qquad a = 1,2, \ldots n-1 \tag{3.54}$$

$$X_\xi = X_i \frac{\partial \chi_i}{\partial \xi} \tag{3.55}$$

Conversely,

$$X_i = \frac{\partial u^a}{\partial \chi i} X_a + \frac{\partial \xi}{\partial \chi i} X_\xi \tag{3.56}$$

Equation (3.44) in conjunction with eq. (3.53) yields

$$X_a du^a = 0 \tag{3.57}$$

where now du^a are completely arbitrary. Thus

$$X_a = 0 \tag{3.58}$$

and therefore, as a consequence of eq. (3.56)

$$X_i = \frac{\partial \xi}{\partial \chi i} X_\xi \tag{3.59}$$

which was the result to be proved.

The orthogonality condition (3.52) ensures that the contravariant components χ_a are also zero since

$$X^a = X_i \frac{\partial u^a}{\partial \chi i} = \frac{\partial \xi}{\partial \chi i} \frac{\partial u^a}{\partial \chi i} = 0 \tag{3.60}$$

Comparison of eq. (3.59) with eq. (3.46) shows that ξ is the entropy and χ_ξ the empirical temperature. Thus the surface

$$T = T(\chi_i) \tag{3.61}$$

is an isentropic surface.

The extension to irreversible systems is straightfoward.

Addendum III :

On the non-integrability of the First Law when q_a vary.

On the basis of eq. (3.26 a,b), eq. (3.27) may be written in the form:

$$\frac{\partial \epsilon}{\partial T} dT + (\frac{\partial \epsilon}{\partial \chi_i} - \chi_i) d\chi_i + \frac{\partial \epsilon}{\partial q_a} dq_a = dQ \tag{3.62}$$

For reversible systems undergoing an adiabatic process, in the absence of heat supply, the following relation exists :

$$T = T(\chi_i) \tag{3.63}$$

One would be tempted to conjecture that, when the system is irreversible, a relation of the type

$$T = T(\chi_i, q_a) \tag{3.64}$$

must also exist. We shall show (a) that eq. (3.64) leads to a contradiction and (b) that is leads to conclusions that are at variance with physical experience.

If eq. (3.64) is valid then the variables T, χ_i and q_a take values that correspond to points on a hypersurface in their space. Therefore eq. (3.62) is integrable and

$$\frac{\partial \epsilon}{\partial T} = \theta \frac{\partial \eta}{\partial T} \tag{3.65a}$$

$$\frac{\partial \epsilon}{\partial \chi_i} - \chi_i = \theta \frac{\partial \eta}{\partial \chi_i} \tag{3.65b}$$

$$\frac{\partial \epsilon}{\partial q_a} = \theta \frac{\partial \eta}{\partial q_a} \tag{3.65c}$$

where, now

$$dQ = \theta d\eta \tag{3.66}$$

However if we introduce the point transformation

$$\theta = \theta(T, \chi_i, q_a) \tag{3.67a}$$

$$\chi_i = \chi_i \tag{3.67b}$$

$$q_a = q_a \tag{3.67c}$$

and in terms of the free energy density ψ, eq.'s (3.65) become

$$\eta = \frac{\partial \psi}{\partial \theta} \tag{3.68a}$$

$$X_i = \frac{\partial \psi}{\partial \chi_i} \tag{3.68b}$$

$$\frac{\partial \psi}{\partial q_a} = 0 \tag{3.68c}$$

where, ϵ and ψ are now functions of θ, χ_i and q_a. But eq. (3.68c) negates the fact that ψ, η and ϵ are functions of q_a, contrary to the original hypothesis. Alternately, if eq. (3.27) is integrable, the system must be reversible. This in turn implies that eq. (3.64) cannot possibly hold. This conclusion can also be corroborated by a physical argument. Imagine a viscoelastic material which is experiencing a homogeneous strain field adiabatically under conditions of relaxation, if the sense that at $t = o$, $\chi_i = o$ but at $t > o^\dagger$, $\chi_i = \chi_i^o$, where χ_i^o are independent of time. In this case, $\dot{\chi}_i = 0$ and $W = 0$ for $t > o^\dagger$. Because the process is adiabatic, $\epsilon = 0$ for $t > o^\dagger$.

Now, in view of eq.'s (3.26) and (3.60) under these conditions

$$\epsilon = \epsilon(\chi_i, q_a) \tag{3.63}$$

Assuming a smooth dependence of ϵ on q_a and χ_i, ϵ being constant for $o^\dagger < t < < \infty$ necessarity implies that q_a must be constant in this interval. But this is contrary to experience, since X_i would then also remain constant, as a result of eq.'s (3.26b) and (3.64). Of course, measurement shows that they don't.

The negation of the relation (3.64) may be shown directly for the case of a thermorheologically simple viscoelastic material undergoing small thermomechanical changes from an equilibrium reference state. It has been shown by the author (15) that under these conditions,

$$a_T h_{i,i} = \frac{\partial}{\partial \xi}\int_o^\xi Cv(\xi - \xi')\frac{\partial \zeta}{\partial \xi'}d\xi' - \theta_o \int_o^\xi a(\xi - \xi')\frac{\partial^2 \epsilon_{kk}}{\partial \xi'}d\xi' + \sum_a \eta_a \dot{q}_a \dot{q}_a \tag{3.69}$$

where a_T is the shift function of temperature and ξ is the reduced time $C_v(\xi)$ is the "specific heat" modulus, $\alpha(\xi)$ is related to the thermal expansion modulus, η_a are constants and ζ is the increment in temperature. Under conditions of zero heat supply and constant strain eq. (3.69) becomes :

$$(3.70) \qquad \frac{\partial}{\partial\xi}\int_0^\xi Cv(\xi-\xi')\frac{\partial\zeta}{\partial\xi'}\,d\xi' + \sum_a \eta_a \dot{q}_a \dot{q}_a = 0$$

It was also shown that q_a are functions of q_a, ζ and ϵ_{kk}. Thus, in the present instance, eq. (3.70) may be written in the form,

$$(3.71) \qquad \frac{\partial}{\partial\xi}\int_0^\xi Cv(\xi-\xi')\frac{\partial\zeta}{\partial\xi'}\,d\xi' + f(q_a{}'\zeta) = 0$$

It is fairly obvious from eq. (3.71) that ζ is not determined simply by the current values of q_a, as eq. (3.64) would have us believe, but depends on the whole history of the internal coordinates q_a.

SECTION 4.

Entropy in the Presence of a Temperature Gradient.

In the previous Section it was established that for a reversible system

$$d\eta = \frac{dQ}{\theta} \tag{4.1}$$

whereas for an irreversible system

$$d\eta > \frac{dQ}{\theta} \tag{4.2}$$

where dQ is the increment of energy supplied to the system by a diathermal process.

The rationale leading to eqs. (4.1) and (4.2) did not in any way depend on whether dQ was supplied by conduction, radiation or heat sources in the system. If, indeed, dQ was supplied solely by conduction then eqs. (4.1) and (4.2) may be written in the form

$$\dot{\eta} + \frac{h^{a}{}_{,a}}{\theta} \geqslant 0 \tag{4.3}$$

the validity of the equality or inequality sign depending on whether the system is reversible or irreversible.

Equation (4.3) can be integrated over the volume V_o of a material region R_χ to give

$$\int_{V_o} \dot{\eta}\, dV + \int_{V_o} \frac{h^{a}{}_{,a}}{\theta}\, dV \geqslant 0 \tag{4.4}$$

After applying the Green-Gauss theorem to eq. (4.4) one obtains

$$\int_{V_o} \dot{\eta}\, aV + \int_{S_o} \left(\frac{h^{a}}{\theta}\right) n_a\, dS + \int_{V_o} \frac{h^{a}\theta_{,a}}{\theta}\, dV \geqslant 0 \tag{4.5}$$

Suppose, now, that the surface S_o is an adiabatic surface, in which case h_i on S vanishes . As a result eq. (4.5) becomes :

$$\int_{V_o} \dot{\eta}\, dV + \int_{V_o} \frac{h^a \theta,_a}{\theta^2}\, dV \geqslant 0 \tag{4.6}$$

At this point we recognize a second thermodynamic inequality. For a choice of a particular thermometer it is possible to define the empirical temperature θ as θ (T). Now provided that $d\theta/dT > 0$, experiments confirm that

$$h^a \theta,_a \leqslant 0 \tag{4.7}$$

i.e., "heat flows from warmer to colder regions"(*). As a result of inequality (4.7) it follows from eq. (4.6) that under adiabatic conditions and for all systems reversible or irreversible,

$$\int_{V_o} \dot{\eta}\, dV \geqslant 0 \tag{4.8}$$

i.e., the entropy of an adiabatically enclosed material system R_χ cannot decrease ; (**) furthermore the equality in eq. (4.8) holds only if the temperature gradient is zero and the system is reversible.

In the presence of heat supply (or absorption) eq. (4.3) becomes :

$$\dot{\eta} + \frac{h_{i,i}}{\theta} - \frac{q}{\theta} \geq 0 \tag{4.9}$$

where the equality sign is to be used when the system is reversible and the inequality sign when the system is irreversible.

(*) Not that inequality (4.7) is separate and distinct from inequality (3,29) which distinguishes between reversible and irreversible systems. In fact inequality (4.7) holds for all systems, reversible or irreversible.

(**) In the absence of heat supply.

Again integrating over the volume V of a region with an adiabatic surface S and as a result of ineq. (4.7) one obtains the "global" inequality

$$\int_{V_o} \dot{\eta} dV \geq \int_{V_o} \frac{\dot{q}}{\theta} dV \tag{4.10}$$

Of course, the sequence of the arguments given above may be reversed. In effect, one may begin with the axiom of non-decreasing entropy of an "isolated" system (one which is devoid of heat sources and sinks and has an adiabatic surface) or, more precisely, with the inequality,

$$\int_{V_o} \dot{\eta} dV \geq \int_{V_o} \dot{\gamma} dV \tag{4.11}$$

where the surface S_0 is adiabatic.

As a result of ineq .(4.11) a sufficient condition that ineq. (4.6) remain inviolate, is that ineq. (4.7) be satisfied. If now, one applies (4.6) and (4.11) to an isolated microsystem (with small V_0) and makes use of the mean value theorem, then a necessary condition that (4.6) remain inviolate in the light of (4.11) is that (4.7) be satisfied.

We conclude, therefore, that ineq. (4.7) is a necessary and sufficient condition for the axiom of non-decreasing entropy of an isolated system.

The Truesdell-Toupin form of the Clausius-Duhem Inequality.

Truesdell and Toupin (7) defined an entropy flux vector $\frac{h^a}{\theta}$. They then asserted that the entropy flux through the surface S_0 of a material region is not greater than the rate of entropy change in the interior of the region i.e.,

$$\int_{V_o} \dot{\eta} dV + \int_{S_o} \frac{h^a}{\theta} n_a dS \geq 0 \tag{4.12}$$

Inequality (4.11) gives rise to the local form

$$\dot{\eta} + \frac{h^{a}{}_{,a}}{\theta} - \frac{h^{a}\theta_{,a}}{\theta^{2}} \geq 0 \tag{4.13}$$

which is their version of the Clausius-Duhem inequality, used extensively in the past by Coleman, Gurtin and others, including the author. However, one must point out that (4.13) is a weaker inequality than (4.3) and (4.7) since these must be satisfied separately. Because of this reason ineq. (4.13) sometimes gives rise to less specific results.

SECTION 5.

Existence of Entropy and Temperature.

In this section we shall show the existence of entropy (and temperature) for reversible and irreversible systems without introducing temperature as a state variable, a priori.

Reversible Systems.

Definition. A system is said to be reversible (a) if under adiabatic conditions its internal energy is a state function of the strains only and (b) if the forces X_i are state functions of the internal energy density ϵ and the strains χ_i.

This definition of a reversible system is in agreement with that given in Section 3 where the existence of temperature was utilized a priori. Evidently, from eq. (3.13a),

$$T = T(\epsilon, \chi_i) \tag{5.1}$$

where upon substitution of eq. (5.1) in eq. (3.13a) yields

$$X_i = X_i(\epsilon, \chi_i) \tag{5.2}$$

in agreement with statement (b) above. Also, for an adiabatic process eq. (3.16) was shown to hold ; this equation in conjunction with eq. (3.13a) yields,

$$\epsilon = \epsilon\,(\chi_i) \tag{5.3}$$

in agreement with statement (a).

The first law now reads,

$$d\epsilon - X_i(\epsilon, \chi_i)\, d\chi_i = dQ \tag{5.4}$$

According to Caratheodory's theorem the Pfaffian form represented by the left hand side of eq. (5.4) is integrable if in the neighborhood of a point P in the space of the variables ϵ and χ_i, there are point P′ which are not accessible from P along curves that are solutions of

$$d\epsilon - X_i(\epsilon, \chi_i)\, d\chi_i = 0 \tag{5.5}$$

However, as a result of our conjecture, as well as a consequence of Addendum II of Section 3 and eq. (5.3), this is so. Thus, under general thermomechanical changes of state,

$$d\epsilon - X_i(\epsilon, \chi_i)\, d\chi_i = \theta(\epsilon, \chi_i)\, d\eta(\epsilon, \chi_i) \tag{5.6}$$

whereupon,

$$\frac{1}{\theta} = \left.\frac{\partial \eta}{\partial \epsilon}\right|_{\chi_i} , \quad X_i = - \theta \left.\frac{\partial \epsilon}{\partial \chi_i}\right|_{\epsilon} . \tag{5.7a, b}$$

Hence, the integrability of the first law, for reversible systems, leads to the existence of two state functions θ and η where,

$$\theta = \theta(\epsilon, \chi_i) , \quad \eta = \eta(\epsilon, \chi_i) . \tag{5.8a, b}$$

If one reverses the roles of θ and ϵ, as dependent and independent variables, in eq. (5.8a) and introduces the free energy density $\psi(\theta,\chi_i)$, then as a result of eq. (5.6),

$$X_i = \frac{\partial \psi}{\partial \chi_i} \tag{5.9a}$$

$$\eta = \frac{\partial \psi}{\partial \theta} . \tag{5.9b}$$

Evidently since eq. (5.9a,b) are identical to the eq.'s (3.23a) and (3.23b) we construe that the functions η and θ are none other than the entropy density and empirical temperature respectively.

<u>Irreversible Systems</u>.

Definition. A system is said to be irreversible if under adiabatic conditions

$$X_i = X_i(\chi_i, \epsilon, q_a) \tag{5.10}$$

but

$$\epsilon \neq \epsilon(\chi_i, q_a). \tag{5.11}$$

The fact that ϵ cannot, under adiabatic conditions, be a state function of χ_i and q_a was demonstrated in Addendum III of Section 3. (Briefly, and at the risk of repeating one's self, if ϵ were a state function of χ_i and q_a then the first law would be integrable in the (ϵ,χ_i,q_a) space with the consequence that $\frac{\partial \psi}{\partial q_a} = 0$, which is a contradiction to the fact that the system is irreversible).

The first law now becomes,

$$d\epsilon - X_i(\epsilon, \chi_i, q_a)\, d\chi_i = dQ. \tag{5.12}$$

Let a point P in the space of the variables ϵ and χ_i represent a thermodynamic state. Then, according to our conjecture, there exist points P′ in the neighborhood P which are not accessible from P by processes that are reversible and adiabatic i.e. along solutions of the equation

$$d\epsilon - X_i(\epsilon, \chi_i, q_a)\, d\chi_i = 0 \tag{5.13}$$

with q_a held constant. Thus by Caratheodory's theorem the left hand side of eq. (5.12) is integrable i.e.

$$d\epsilon - X_i(\epsilon, \chi_i, q_a)\, d\chi_i = \theta(\epsilon, \chi_i, q_a)\, d\eta(\epsilon, \chi_i, q_a) \Big|_{q_a = \text{const.}} \tag{5.14}$$

Equation (5.14) warrants the following statement :

<u>For irreversible systems, with q_a constant, the First Law is integrable. Hence for such systems there exist state functions θ, called temperature, and η, called entropy, such that</u>

$$\theta = \theta(\epsilon, \chi_i, q_a), \eta = \eta(\epsilon, \chi_i, q_a).$$

By interchanging the roles of ϵ and θ as dependent and independent variables one finds that

$$\left.\frac{\partial \epsilon}{\partial \theta}\right|_{q_a} d\theta + \left.\frac{\partial \epsilon}{\partial \chi_i}\right|_{q_a} d\chi_i - X_i\, d\chi_i = \theta\, d\eta \Big|_{q_a} \tag{5.15}$$

which is identical to eq. (3.25) of Section 3. Thus the conclusions of Section 3 follow.

SECTION 6.

A Mechanical Model for the Internal Variables.

In the case of reversible thermodynamics the perfect gas is a suitable physical system which is customarily used to explain the physical connotation of the mathematics involved.

In the case of irreversible thermodynamics a very suitable system is a mechanical system consisting of elastic springs and viscous dashpots in some geometric array and immersed in a heat bath ; the heat bath may be visualized as a perfect gas offering no resistance to the motion of the springs and dashpots but having a significant heat capacity.

Such a system is shown in the following Figure :

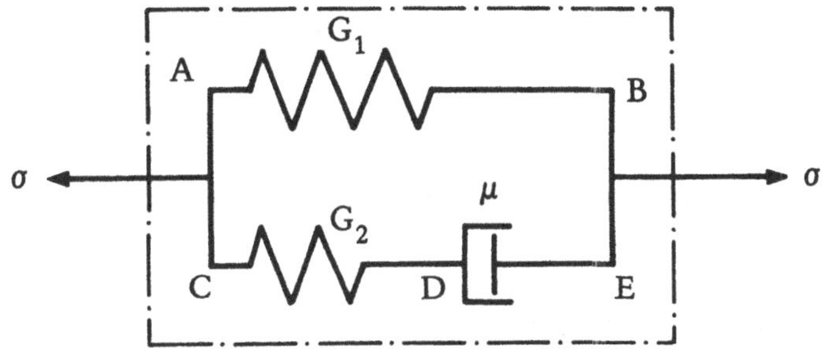

The system consists of a spring in series with a dashpot and these are connected in parallel with another spring. The assembly is immersed in a heat bath which has an infinitely flexible adiabatic external wall.

If we were to regard the system as a "black box" in the sense that we knew nothing about its internal structure we would soon find that, if the system were subjected to a stress σ, the state of stress could not be described uniquely in terms of the "overall strain" (say extended length over initial length minus unity) of the system and its temperature. Thus our system belongs to the class of irreversible systems.

However if we were to open the system and become aware of its internal structure or constitution we would soon realize that σ could be described uniquely if in addition we knew the strain of the dashpot or the position of the point D relative to point E. The position of the point D relative to E essentially constitutes an internal coordinate in the sense that it is not macroscopically or phenomenologically observable.

Let us assume that the heat bath is large with a large heat capacity so that changes in temperature due to deformation of the system are negligible. Under isothermal conditions eqs. (3.36) and (3.39) give the following relation

$$X_i \dot{x}_i = \psi + \theta\dot{\gamma} \tag{6.1}$$

Since ψ has the significance of recoverable energy, $\theta\dot{\gamma}$ must signify the rate of energy necessary to overcome internal dissipative resistance which in the case of the above system must be rate of work necessary to extend the dashpot.

Let q be the strain of the dashpot and μ its velocity coefficient. The force applied to the dashpot is then $\mu\dot{q}$ and the rate of work done on the dashpot in $\mu\dot{q}^2$. Thus

$$\theta\dot{\gamma} = \mu\dot{q}^2 \tag{6.2}$$

It follows that since ψ is the recoverable energy it must be the elastic energy stored in the springs in which case,

$$\psi = \frac{1}{2} G_1 \, \epsilon^2 + \frac{1}{2} G_2 (\epsilon - q)^2 \tag{6.3}$$

where ϵ is the "observable" strain of the system. Also as a result of eqs. (6.2) and (3.39),

$$\frac{\partial\psi}{\partial q} + \mu\dot{q} = 0 \tag{6.4}$$

whereas eq. (3.23a) yields,

$$\sigma = \frac{\partial\psi}{\partial\epsilon} \tag{6.5}$$

Use of eq. (6.3) in eqs. (6.4) and (6.5) gives the following relations for σ and q:

$$\sigma = \epsilon\, G_1 + (\epsilon - q)\, G_2 \tag{6.6}$$

$$\mu\dot{q} + G_2 q = \epsilon\, G_2 \tag{6.7}$$

Letting $\lambda = \frac{G_2}{\mu}$ and assuming that at $t = -\infty$ the system is undisturbed in the sense that $\sigma = \epsilon = q = 0$, eq. (6.6) and (6.7) may be solved easily to give

$$\sigma(t) = \epsilon(t)\, G_1 + G_2 \int_{-\infty}^{t} e^{-\lambda(t-\tau)} \frac{\partial \epsilon}{\partial \tau}\, d\tau \tag{6.8}$$

It may be verified that this relation between σ and ϵ may be obtained by application of purely mechanical principles without invoking thermodynamics. It is gratifying that both approaches give the same result.

The case of a spring and dashpot in parallel can be obtained from eq. (6.8) by a limiting process in which $G_2 \to \infty$, i.e., the spring becomes infinitely stiff. In this event eq. (6.8) becomes

$$\sigma(t) = \epsilon(t)\, G_1 + \mu \lim_{\lambda \to \infty} \int_{-\infty}^{t} \lambda\, e^{-\lambda(t-\tau)} \frac{\partial \tau}{\partial \tau} d\tau \tag{6.9}$$

However

$$\lim_{\lambda \to \infty} \lambda\, e^{-\lambda(t-\tau)} = \delta\,(t-\tau) \tag{6.10}$$

where $\delta(t)$ is the Dirac δ-function. Hence as a result of eq. (6.10) eq. (6.9) takes the more familiar form

$$\sigma(t) = G_1\, \epsilon(t) + \mu\dot{\epsilon}(t) \tag{6.11}$$

The application of the theory to more elaborate spring-dashpot systems in parallel is straightforward. One such system is shown below

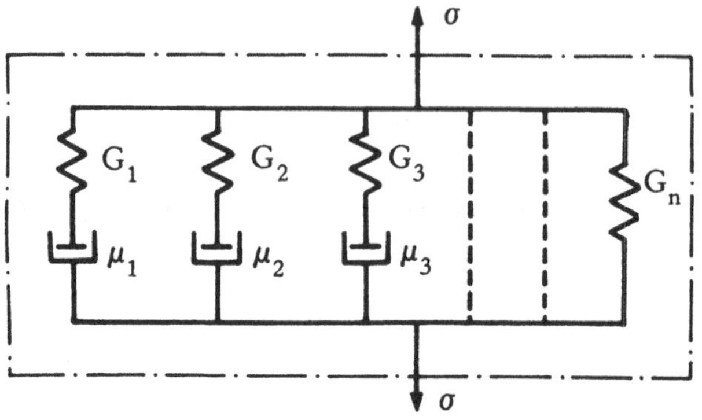

Again let q_1 be the strain in dashpot 1, etc. Then under isothermal conditions

(6.12) $$\psi = \frac{1}{2}\epsilon^2 G_n + \frac{1}{2}\sum_{a=1}^{n-1} (\epsilon - q_a)^2 G_a,$$

(6.13) $$\theta\dot{\gamma} = \sum_a \mu_a \dot{q}_a^{\,2} ,$$

(6.14) $$\frac{\partial\psi}{\partial q_a} + \mu_a \dot{q}_a = 0$$

(a not summed)

and

(6.15) $$\sigma = \frac{\partial\psi}{\partial\epsilon} .$$

As a result of eq. (6.15)

(6.16) $$\sigma(t) = \epsilon(t)\, G_n + \sum_{a=1}^{n-1} (\epsilon(t) - q_a(t))\, G_a$$

whereas eqs. (6.12) and (6.14) give :

(6.17) $$\mu_a \dot{q}_a + G_a q_a = \epsilon\, G_a$$

(a not summed)

Again let $\sigma = \epsilon = q_a = 0$ at $t = -\infty$.

Then integration of eq. (6.17) yields :

$$\epsilon - q_a = \int_{-\infty}^{t} e^{-\lambda_a(t-\tau)} \frac{\partial \epsilon}{\partial \tau} d\tau \tag{6.18}$$

where $\lambda_a = (\frac{G_a}{\mu_a})$. As a result, eq. (6.16) gives σ in terms of the history of $\epsilon(\tau)$, i.e.,

$$\sigma(t) = G_n\, \epsilon(t) + \sum_{a=1}^{n-1} G_a \int_{-\infty}^{t} e^{-\lambda_a(t-\tau)} \frac{\partial \epsilon}{\partial \tau} d\tau \tag{6.19}$$

which again is identical to the expression that could have been obtained from purely mechanical considerations.

Interestingly enough the mechanical response of a generalized "Kelvin model" shown below can also be derived in a straightforward fashion, from thermodynamic considerations.

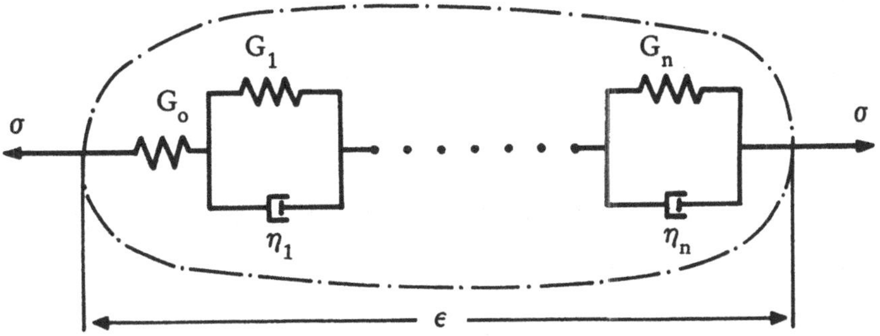

In this model the strain ϵ is expressed in terms of a hereditary integral involving the stress history, i.e., the situation is reversed insofar as the stress plays the role of the independent variable whereas the strain becomes a state variable.

In terms of the general problem χ_i become the dependent variables and χ_i the independent variables in eq. (3.25) which may be written in the form

$$\dot{\epsilon}\Big|_{q_r} - X_i \dot{\chi}_i = \theta \dot{\eta}\Big|_{q_r} \tag{6.20}$$

where

$$\phi = \psi - \chi_i X_i \tag{6.21}$$

As a result of eq.'s (6.20) and (6.21) one finds that

$$\eta = -\frac{\partial \phi}{\partial \theta} \tag{6.22}$$

and

$$\chi_i = -\frac{\partial \phi}{\partial X_i} \tag{6.23}$$

where now

$$\phi = \phi(X_i, q_r) \tag{6.24}$$

Furthermore, under isothermal conditions, and utilizing eq.'s (6.1), (6.21), and (6.23) one obtains the result :

$$\frac{\partial \phi}{\partial q_r} \dot{q}_r + \theta \dot{\gamma} = 0 \tag{6.25}$$

where q_r are still the strains in the dashpots.

Again since

$$\theta \dot{\gamma} = \sum_{r=1}^{n} \eta_r \dot{q}_r \dot{q}_r, \tag{6.26}$$

eq. (6.25) is satisfied if

$$\frac{\partial \phi}{\partial q_r} + \eta_r \dot{q}_r = 0 \tag{6.27}$$

(r not summed)

It is easily shown that eq.'s (6.23) and (6.27) completely describe the mechanical behavior of the model. In effect,

$$\phi = -\frac{\sigma^2}{2G_o} + \sum_{r=1}^{n} \frac{1}{2} G_r q_r^2 - \sigma \sum_{r=1}^{n} q_r \tag{6.28}$$

In view of eq.'s (6.23) and (6.28)

$$\epsilon = \frac{\sigma}{G_o} + \sum_{r=1}^{n} q_r \tag{6.29}$$

whereas eq.'s (6.27) and (6.28) yield

$$G_r q_r + \eta_r \dot{q}_r = \sigma \tag{6.30}$$

(r not summed)

Equations (6.29) and (6.30) are indeed the equations of compatibility and equilibrium of the model and together they give the required result :

$$\epsilon(t) = \frac{\sigma}{G_o} + \sum_{r=1}^{n} \frac{1}{\eta_r} \int_{-\infty}^{t} e^{\lambda_r (t-\tau)} \sigma(\tau)\, d\tau \tag{6.31}$$

where

$$\lambda_r = \left(\frac{G_r}{\eta_r}\right). \tag{6.32}$$

SECTION 7.

On the physical nature of the internal variables.

We begin by recapitulating the main results derived above :

(7.1) $$\psi = \psi(\theta, \chi_i, q_a)$$

(7.2) $$X_i = \frac{\partial \psi}{\partial \chi_i}$$

(7.3) $$\eta = -\frac{\partial \psi}{\partial \theta}$$

(7.4a) $$\frac{\partial \psi}{\partial q_a} \dot{q}_a + \theta \dot{\gamma} = 0$$

(7.4b) $$\frac{\partial \psi}{\partial q_a} \dot{q}_a \leq 0$$

To these we add the heat conduction inequality :

(7.5) $$-h^a \theta_a \geq 0$$

Recall that eq.'s (7.2) and (7.3) were derived following the assumed existence of the internal variables, the axiom of inaccessibility of thermodynamic states by processes that are reversible and adiabatic and the consequent existence of the state function "entropy".

In this Section we shall show that eq.'s (7.1 - 7.4) do in fact have a physical basis and completely correspond to equations which are derived by using statistical thermodynamics for the treatment for the motion of coiling polymer molecules, under applied stress fields. Furthermore the study of this model will enable us to generalize and extend the theory without the danger of introducing physically vacuous concepts.

At this juncture it is well to emphasize that q_a need not be scalar, but could denote the components of a tensor of any order, provided they satisfy certain invariance requirements associated with the motion of the spatial frame. This point will be discussed at length in later Sections. In the discussion which immediately follows, q_a are the components of vectors.

Motion of a polymer molecule.

The kinetic theory of elasticity of rubber consists in finding the probability $p(y_1, y_2, y_3)dV$ that one end of a molecule will be in a volume element dV in the neighborhood of a point (y_1, y_2, y_3) in a cartesian frame of reference y_i when the other end is fixed at the origin of the coordinates (18). In the above theory an approximate value for p is found by assuming that the molecule is a chain of n freely jointed links each of length a ; in this event in the limiting case when $n \to \infty$ and $a \to 0$ such that na^2 remains finite,

$$p = \left(\frac{3}{2\ell^2 \pi}\right)^{3/2} \exp \left\{-\frac{3}{2\ell^2} y_i y_i\right\} \tag{7.6}$$

where ℓ is the root mean square length of the molecule, i.e.

$$\ell^2 = na^2 \tag{7.7}$$

If the molecule is free to move without external constraints it will eventually assume an equilibrium configuration which maximizes p i.e. the free end will also lie at the origin $y_i = 0$.

The statistical definition of entropy η is :

$$\eta = -k \log p \tag{7.8}$$

where k is Bolzmann's constant.

The fact that the molecule is regarded as a free linkage implies that its internal energy is zero. Thus the force F_i that must be exerted to keep energy ψ of the molecule is therefore given by

(7.9) $$\psi = - k\theta \log p$$

in the usual notation, or

(7.10) $$\psi = -\frac{3}{2} k\theta \log\left(\frac{3}{2\ell^2\pi}\right) + \frac{3}{2} k\theta \frac{y_i y_i}{\ell^2}$$

Thus, since

(7.11) $$F_i = \psi_{,i}$$

it follows that

(7.12) $$F_i = 3k\theta \frac{y_i}{\ell^2}$$

Thus the molecule behaves as an entropic linear spring where the force constant of the spring is $\frac{3k\theta}{\ell^2}$.

Evidently, if the fixed end of the molecule does not lie at $y_i = 0$, but at some point y_i°, then eq. (7.6) becomes :

(7.13) $$p = \left(\frac{3}{2\ell^2\pi}\right)^{3/2} \exp\left\{-\frac{3}{2\ell^2}\sum_{i=1}^{3}(y_i - y_i^0)(y_i - y_i^0)\right\}$$

whereas eq. (7.12) now reads,

(7.14) $$F_i = \frac{3k\theta}{\ell^2}(y_i - y_i^0).$$

The significance of Rouse's work (19) lies in the recognition that if a molecule is sufficiently long, it may be divided into a finite number of sub-molecules each of which may be regarded as a chain of n freely jointed links. Thus, if a sub-molecule is not in its equilibrium configuration it will have to endure a force, which is supplied by the interaction of the sub-molecule with the medium in which it lies.

This means that the molecule, (as a result of its interaction with its immediate milieu), does not simply experience a force at its free end, but a distribution of forces all along its length. We shall call such a distribution of forces a force field. Let there be N sub-molecules and let r denote the junction between the (r-1)' th and the r'th sub-molecule (the loose ends of the first and last molecules are also considered as junctions). Let the configuration of the molecule be defined by the set of coordinates $y_i^{(r)}$ of the junctions r.

The probability density corresponding to a certain configuration of the molecule is equal to the product of the probability densities of the sub-molecules which make up that particular configuration.

Thus for the complete molecule

$$p = \left(\frac{3}{2\ell^2\pi}\right)^{3N/2} \exp\left\{-\frac{3}{2\ell^2}\sum_{r=1}^{N+1}(y_{r+1}-y_r)^2\right\} \tag{7.15}$$

whereby it follows that the free energy of the molecule becomes :

$$\psi = -\frac{3N}{2}k\theta \log\left(\frac{3}{2\ell^2\pi}\right) + \frac{3}{2}\frac{k\theta}{\ell^2}\sum_{r=1}^{N+1}(y_{r+1}-y_r)^2 \tag{7.16}$$

Let $G_i^{(r)}$ be the force vector at the junction r, then as a result of eq. (7.11),

$$G_i^{(r)} = \frac{\partial\psi}{\partial y_i^{(r)}} \tag{7.17}$$

Note that

$$\eta = -k \log p = -\frac{\partial\psi}{\partial\theta} \tag{7.18}$$

In what follows we shall limit ourselves to isothermal motions. Let $y^{\circ(r)}$ denote an initial configuration of a molecule under the initial force field $G_i^{(r)}$. The molecule is now disturbed and takes on a new configuration $y_i^{(r)}$. Let $g_i^{(r)}$ be the additional force field which is necessary to maintain the molecule in its disturbed configuration. Set

$$u_i^{(r)} = y_i^{(r)} - y_i^{o(r)} \tag{7.19}$$

and let $\Delta\psi$ be the change in free energy, where as a result of eq. (7.16)

$$\Delta\psi = \frac{3}{2}\frac{k\theta}{\ell^2}\sum_{r=1}^{N+1}(u_i^{(r+1)} - u_i^{(r)})(u_i^{(r+1)} - u_i^{(r)}) \tag{7.20}$$

(i summed, i = 1,2,3)

Then as a result of eq.'s (7.16) and (7.17)

$$g_i^{(r)} = \frac{\partial\Delta\psi}{\partial u_i^{(r)}} \tag{7.21}$$

As it happens, eq. (7.20) may be written in the form

$$\Delta\psi = \frac{3}{2}\frac{k\theta}{\ell^2} A^{rs}\, u_i^{(r)}u_i^{(s)} \tag{7.22}$$

where

$$A^{rs} = \begin{bmatrix} 1 & -1 & & & \\ -1 & 2 & -1 & & \\ & -1 & 2 & -1 & \\ & & -1 & 2 & -1 \\ & & & -1 & 1 \end{bmatrix} \tag{7.23}$$

In the following figure AB is the "macroscopic" face of a specimen (shown in part) which is to be tested by applying stress on the face AB. Slightly protruding from this face is a molecule OP, which experience a force f_i at O, in the y_i direction, as a result of the stress on AB.

At point interior to the region ABCD the molecule will experience forces g_i which are the result of the interaction of the molecule with its immediate milieu. However at 0,

$$g_i = f_i \tag{7.24}$$

Since the specimen cannot be made smaller without negating the concept of a continuum, then the point 0 is the only part of the molecule where force and/or displacement can be monitored. In this sence the displacement at 0 $(r = 1)$ is an observed variable (which we shall simply denote by u_i) whereas the displacements of points of the molecule interior to the speciment $(1 < r < n+1)$, are internal variables ; these shall be denoted by $q_i^{(r)}$ where

$$q_i^{(r-1)} = u_i^{(r)} \quad (r = 2, 3 \ldots n + 1) \tag{7.25}$$

Since the forces $g_i^{(r)}$ $(1 < r)$ are resistive they will tend to oppose the motion of the molecule in which case

$$g_i^{(r)}\dot{q}_i^{(r)} \leq 0 \text{ (r not summed)} \tag{7.26}$$

The equality being valid if and only if $g_i^{(r)}$ and/or $q_i^{(r)}$ are zero. Of course

$$g_i^{(r)} = \frac{\partial \Delta\psi}{\partial q_i^{(r)}} \tag{7.27}$$

It follows from eq. (7.24) that

$$F_i = \frac{\partial \Delta\psi}{\partial u_i} \tag{7.28}$$

whereas from eq.'s (7.26) and (7.27)

$$\sum_{r=1}^{N+1} \frac{\partial \Delta\psi}{\partial q_i^{(r)}} \dot{q}_i^{(r)} \leq 0 \text{ (i summed)} \tag{7.29}$$

Of course the left hand side of the inequality (7.29) is nothing else but the rate of energy dissipation in the medium, as a result of the motion of the molecule.

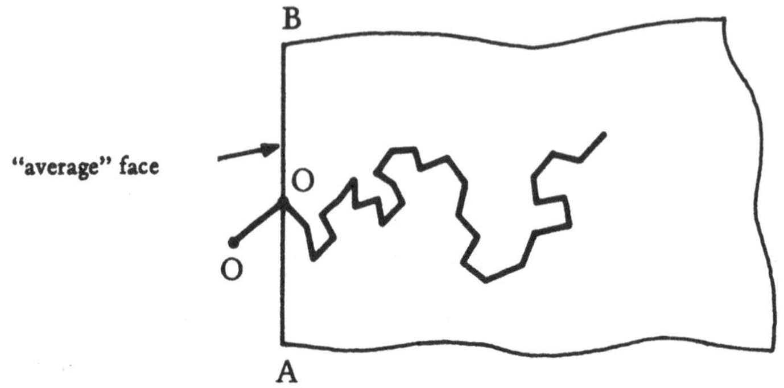

Equations (7.28), (7.18) and (7.29) are indeed exactly analogous to the continuum equation (7.2), (7.3) and (7.4b).

Evidently, for the molecule

$$\theta\dot{\gamma} = \sum_{r=1}^{N+1} - \frac{\partial \Delta\psi}{\partial q_i^{(r)}} \dot{q}_i^{(r)} \geq 0 \tag{7.30}$$

<u>Determination of the motion of the molecule.</u>

The question that poses itself at this time is :

"Given the history $f_i(\tau)$, $(0 < \tau < t)$, where t is the present time, can one determine the motion of the entire molecule, and in particular $u_i(t)$; conversely, given $u_i(t)$ can one determine the motion of the entire molecule and in particular $f_i(t)$? ".

A persual of eq.'s (7.27) and (7.28) shows that this cannot be done, unless the force $g_i^{(r)}$ are related to the motion of the molecule. This implies a constitutive description of the material which is "finer" or more detailed than a purely phenomenological description.

As a first step let us assume that the molecule views its immediate milieu as a Newtonian fluid. In this event

$$g_i^{(r)} = - \eta \, \dot{q}_i^{(r)} \tag{7.31}$$

where η is the viscocity coefficient to the fluid. Equation (7.31), which we shall call INTERNAL CONSTITUTIVE EQUATION, provides the missing constitutive relation, which is sufficient for the unique determination of the motion of the molecule.

The equations of motion of the molecule are, therefore :

$$f_i = \frac{\partial \Delta \psi}{\partial u_i} \tag{7.32}$$

$$\frac{\partial \psi}{\partial q_i^{(r)}} + \eta \, \dot{q}_i^{(r)} = 0 \tag{7.33}$$

where eq. (7.33) has been obtained by combining eq.'s (7.27) and (7.31).

Also as a result of eq.'s (7.27), (7.29) and (7.31) and the definition of irreversible entropy it follows that

$$\theta \dot{\gamma} = \sum_{r=1}^{N+1} \eta \, \dot{q}_i^{(r)} \dot{q}_i^{(r)} \geq 0 \tag{7.34}$$

The inequality is satisfied if η is positive.

Equation (7.33) was inferred from purely phenomenological considerations by the author in Ref.'s 13, 15 and 16.

Solution of the eq.'s of motion.

We shall now obtain an explicit solution for $f_i(t)$ given $u_i(t)$ (or vice-versa), by solving eq.'s (7.32) and (7.33). For simplicity we shall assume that the molecule undergoes unidirectional motion i.e. u_i (and $q_i^{(r)}$) are different from zero only for $i = 1$.

Thus setting $f_1 \equiv f$, $u \equiv q^{\circ}$, $q_1^{(r)} \equiv q^r$ ($r = 1, 2 \ldots . N + 1$) and as a result of eq.'s (4.22), (4.23), (4.32) and (4.33) we have the following set of eq.'s :

$$\frac{3k\theta}{\ell^2}(q^0 - q^1) = f(t) \tag{7.35a}$$

(7.35b)
$$\frac{3k\theta}{\ell^2}\left(-q^{r-1} + 2q_r - q^{r+1}\right) + \eta\frac{dq^r}{dt} = 0$$
$$r = 1, 2 \ldots\ldots N$$

(7.35c)
$$\frac{3k\theta}{\ell^2}\left(-q^N + q^{N+1}\right) + \eta\frac{dq^{N+1}}{dt} = 0$$

where it has been assumed that the root mean square lengths of the sub-molecules are equal to ℓ and that the end of the molecule at $r = N + 1$ is allowed to drift. On the other hand if this end of the molecule is fixed then $q^{N+1} = 0$, and eq. (7.35c) reads instead :

(7.35d)
$$-\frac{3k\theta}{\ell^2}\, q^N = {}^{N+1}(t)$$

where f^{N+1} is the force at $r = N + 1$ necessary to keep the molecule fixed at the point. The above eq.'s are subject to the initial condition $q^r = 0$ ($r = 0, 1, 2 \ldots \ldots N + 1$) for $t < 0$.

We now seek a solution when $q^{N+1} \equiv 0$, and f(t) is given. Under these conditions the above equations can be cast in the matrix form,

(7.36)
$$[a]\,\{q\} + a\,\{\dot{q}\} = \{f(t)\}\frac{\ell^2}{3k\theta}$$

where

$\{q\}$ is the column vector $\{q_1, \ldots\ldots, q_N\}$, $\{f\}$ is the column vector $\{f(t), 0, 0 \ldots\ldots, 0\}$ and $[a]$ is the square, positive definite non-singular matrix :

(7.37)
$$[a] = \begin{bmatrix} 2 & -1 & & & & \\ -2 & 2 & -1 & & & \\ & -1 & 2 & -1 & & \\ & \cdots & \cdots & \cdots & \cdots & \cdots \\ & & & -1 & 2 & -1 \\ & & & & -1 & 2 \end{bmatrix}$$

and $\alpha = \dfrac{\eta \ell^2}{3k\theta}$.

Of course relation (7.36) represents a system of N linear first order ordinary differential equations with constant coefficients. These we solve as follows. Because [a] is positive definite and symmetric there exists an orthonormal matrix [R] which diagonalizes [a] in the sence that

$$R_{sn}\ R_{tm}\ a_{nm} = \delta_{st}\ a_t \qquad (7.38)$$

where δ_{st} is the Kronecker delta a_t (t = 1, 2 . . . N) are the eigenvalues of [a].

Let

$$[R]\ \{q\} = \{p\},\ \{q\} = [R]^T\ \{p\} \qquad (7.39a,b)$$

Then, as a result of eq.'s (7.36), (7.38) and (7.39) and in view of the stipulated conditions we obtain the uncoupled set of equations :

$$a_n p_n + \alpha \dot{p}_n = R_{n1}\ f(t) \frac{\ell^2}{3k\theta} \qquad (7.40)$$

These may be integrated immediately using Laplace Transform (or otherwise) to yield

$$p_n = \frac{R_{n1}}{\eta} \int_0^t e^{-a_n(t-\tau)/\alpha}\ f(\tau)\ d\tau \qquad (7.41)$$

It follows that

$$q_1 = \sum_{n=1}^{N} \frac{R^2_{n1}}{\eta} \int_0^t e^{-a_n(t-\tau)/\alpha}\ f(\tau) d\tau \qquad (7.42)$$

If the molecule does not protrude through the face AB then $00' = o$, i.e., $\ell_{00'} = 0$, in which case, from eq. (7.35), $q_1 = u$, i.e.,

(7.43) $$u(t) = \sum_{n=1}^{N} \frac{R^2_{n1}}{\eta} \int_o^t e^{-a_n(t-\tau)/a} f(\tau) d\tau$$

Note :

(7.44) $$R_{nm} = \sin \frac{nm\pi}{(N+1)}$$

(7.45) $$a_n = 4 \sin^2 \frac{n\pi}{2(N+1)}$$

If we now consider m molecules to emanate from the face AB per unit area and f_s to be the force on molecule s, then the stress σ on the face is simply

(7.46) $$\sigma = \sum_{s=1}^{m} f_s$$

Furthermore if the test specimen is of length L, the phenomenological strain ϵ in the specimen is simply $\frac{u}{L}$. Thus in terms of ϵ and σ and in view of eq. (7.46) eq. (7.43) becomes :

(7.47) $$\epsilon = \sum_{n=1}^{N} \frac{R^2_{n1}}{\eta L} \int_o^t e^{-a_n(t-\tau)/a} \sigma(\tau) d\tau$$

or

(7.48) $$\epsilon = \int_o^t J(t-\tau) \frac{\partial \sigma}{\partial \tau} d\tau$$

where the creep function J(t) can be found from (7.47) by putting $\sigma(\tau) = H(\tau)$ where $H(\tau)$ is the Heaviside step function. It follows that

(7.49) $$J(t) = \sum_{n=1}^{N} \frac{R^2_{n1}}{3k\theta L} \{1 - e^{-a_n t/a}\}$$

Some very important conclusions arise from the study of this idealized material, which substantiate the points that we made in the Introduction.

From eq.(7.43) we see that the observable u(t),cannot be determined uniquely from the value of the state function at t, i.e., f(t), or conversely f(t) is not uniquely determined by u(t), so long as the internal material points of the system are in motion. Similarly persual of eq. (7.20) shows that $\psi(t)$ is not uniquely determined by u(t), but $q_r(t)$ r = 1, 2, . . . are also necessary for this purpose. Thus, an equation of state which provides a one to one correspondence the state functions and the observable variables does not exist.

However such an equation does exist, when one includes in the set of state variables, the displacement of internal material points, as well as an internal constitutive equation (eq. (7.33)), which makes these internal displacements determinate.

On the other hand as the material tends toward an equilibrium state, i.e., as

$$f(t) \to f_\infty H(t) \tag{7.50}$$

then from eq. (7.43)

$$u_\infty = f_\infty \sum_{n=1}^{N} \frac{Rn^2{}_1}{3k\theta} \tag{7.51}$$

i.e., in the equilibrium state there is an one-to-one relation between the value of f and the values of u and θ.

Similarly, since in the equilibrium state $\{q\} = \{0\}$, and as a result of eq.'s (7.36) and (7.51), then $\{q\}_\infty$ are uniquely related to u_∞. Hence persual of eq. (7.20) shows that ψ_∞ is uniquely determined by the equilibrium value of the variable u and the equilibrium value of the temperature, as stated in the Introduction.

SECTION 8.

Thermomechanical Constitutive Equations of Elastic Materials.

These equations, being by far the easiest to derive, provide a basis for comparison of Coleman's method of approach (using the Truesdell-Toupin form of the form of the Clausius-Duhem inequality) and our approach which utilizes directly the results derived in Section 3. We begin with Coleman's approach which we discuss below.

In Section 1 it was pointed out that the deformation of a material region R_χ is defined with respect to a cartesian spatial frame of reference y_i by means of the triad of relations

$$y_i = y_i(\chi^a, t) \tag{8.1}$$

In this frame the first law and the Truesdell-Toupin form of the Clausius-Duhem inequality take the forms :

$$\rho\dot{\epsilon} = T_{ij}\, v_{i,j} - h_{i,i} + \rho q \tag{8.2}$$

$$0 \leq T_{ij}\, v_{i,j} + \rho(\theta\dot{\eta} - \dot{\epsilon}) - \frac{1}{\theta} h_i\, \theta_{,i} \tag{8.3}$$

Note that the frame of reference y_i is completely arbitrary and in fact any other frame $\bar{y}_i$, which differs from the original one by a rigid body translation and/or rotation, i.e.,

$$\bar{y}_i = Q_{ij}(t) y_j + C_i(t) \tag{8.4}$$

where

$$Q_{ik}(t)\, Q_{jk}(t) = \delta_{ij} = Q_{ki}(t)\, Q_{ji}(t), \tag{8.5}$$

is admissible. Equations (8.2) and (8.3), being scalar in character, remain form-invariant with a rigid body transformation (eq. (8.4)) of the original frame.

However, eqs. (8.2) and (8.3) present a difficulty ; for instance, if one compares eq. (8.2) with eq. (1.16) one finds that $v_{i,j}$ are not material derivatives of any particular quantity. Nonetheless one can show that

$$v_{i,j} = \frac{\partial \chi^a}{\partial y_j} \frac{\partial v_i}{\partial \chi^a} = \frac{\partial \chi^a}{\partial y_j} \left(\frac{\partial y_i}{\partial \chi_a} \right)^{\cdot} . \tag{8.6}$$

in which case eqs. (8.2) and (8.3) can be written in the alternative but "thermodynamically" canonical form

$$\rho \dot{\epsilon} = T^a_{\ i} \left(\frac{\partial y_i}{\partial \chi^a} \right)^{\cdot} - h_{i,i} + \rho \dot{q} \tag{8.7}$$

$$T^a_{\ i} \left(\frac{\partial y_i}{\partial \chi^a} \right)^{\cdot} + \rho(\theta\dot{\eta} - \dot{\epsilon}) - \frac{1}{\theta} h_i \, \theta_{,i} \geq 0 \tag{8.8}$$

where $\dfrac{\partial y_i}{\partial \chi^\alpha} \equiv y_{i,\alpha}$ is the deformation gradient tensor and

$$T^a_{\ i} = \frac{\partial \chi^a}{\partial y_j} T_{ij} \tag{8.9}$$

Equation (8.8) can be expressed in terms of the free energy density ψ (per unit mass). In terms of ψ,

$$T^a_{\ i} \left(\frac{\partial y_i}{\partial \chi^a} \right) - \rho(\dot{\psi} + \eta\dot{\theta}) - \frac{1}{\theta} h_i \, \theta_{,i} \geq 0 \tag{8.10}$$

where

$$\psi = \epsilon - \eta\theta \tag{8.11}$$

To these equations one must add the equation of conservation of mass,

$$\left| \frac{\partial y_i}{\partial \chi^a} \right| = \frac{\rho_o}{\rho} \tag{8.12}$$

and the equation of conservation of linear momentum

(8.13) $$T^{\alpha}_{i,\alpha} + \rho f_i = \frac{Dv_t}{Dt}$$

and the equation of conservation of angular momentum

(8.14) $$T^{\alpha}_{i} \frac{\partial x^{\beta}}{\partial y_i} = T^{\beta}_{i} \frac{\partial x^{\alpha}}{\partial y_i}$$

The whole thermomechanical process is described by the following set of variables, twenty in number :

$$\rho, T^{\alpha}_{i}, \eta, \epsilon, \psi, h_i, y_i \text{ and } \theta.$$

These variables are related by the eq's. (8.7), (8.11), (8.12), (8.13) and (8.14), which constitute nine relations between the variables. In addition the relations between the variables must satisfy the constraint imposed by the inequality (8.10).

The implication is that the response of an elastic material cannot be depicted from the "field equations" alone, but what is needed in addition is a set of constitutive equations (equations of state) relating the state functions T^{α}_{i}, η, ϵ, ψ and h_i to the state variables $y_{i.\alpha}$ and θ.

By definition, the state functions of an elastic material depend on the current values of the state variables (possibly their current spatial gradients) but not on their past values. Thus the equations of state will be of the form :

(8.15a) $$T^{\alpha}_{i} = T^{\alpha}_{i} (y_{i,\alpha}, \theta, \theta_{,i})$$

(8.15b) $$\eta = \eta (y_{i,\alpha}, \theta, \theta_{,i})$$

(8.15c) $$\epsilon = \epsilon (y_{i,\alpha}, \theta, \theta_{,i})$$

(8.15d) $$\psi = \psi (y_{i,\alpha}, \theta, \theta_{,i})$$

(8.15e) $$h_i = h_i (y_{i,\alpha}, \theta, \theta_{,i})$$

where we have included dependence on $\theta_{,i}$ as a possibility. Equation (8.15) provide fifteen relations, however, of these only eleven can be independent since the state functions must also satisfy relations (8.11) and (8.14). These are the eleven missing relations between the state functions and the state variables. Substitution of eqs. (8.15) in eq. (8.10) yields the result

$$\left\{T^{a}{}_{i} - \rho \frac{\partial \psi}{\partial y_{i,a}}\right\} \left(\frac{\partial y_i}{\partial \chi^a}\right)^{\bullet} - \rho \frac{\partial \psi}{\partial \theta} + \eta \dot{\theta} - \rho \frac{\partial \psi}{\partial \theta_{,i}} \dot{\theta}_{,i} - \frac{1}{\theta} h_i \dot{\theta}_{,i} \geq 0 \tag{8.16}$$

One can verify that all deformation histories $y_i(\chi^a, \tau)$ and temperature histories $\theta(\chi^a,\tau)$ are admissible, since given any such histories (which determine the state functions uniquely, as a result of eqs. (8.15)) body force and heat source densities f_i and q can always be found, such that eqs. (8.7) and (8.12) are satisfied. Evidently, since all deformation and temperature histories are admissible, the deformation gradient rates $\dfrac{\partial y_i}{\partial \chi^a}$ and θ and $\theta_{,i}$ can be chosen at will ; furthermore inequality (8.16) must not be violated for all such choices. This is possible if and only if eqs.

$$T^{a}{}_{i} = \rho \frac{\partial \psi}{\partial y_{i,a}} \tag{8.17a}$$

$$\eta = - \frac{\partial \psi}{\partial \theta} \tag{8.17b}$$

$$h_i = h_i (y_{i,a}, \theta, \theta_{,i}) \tag{8.17c}$$

$$\frac{\partial \psi}{\partial \theta_{,i}} = 0 \tag{8.17d}$$

$$-\frac{1}{\theta} h_i \theta_{,i} \geq 0 \tag{8.18}$$

are verified.

Eq. (8.20) does in fact show that the state variables do not depend on the temperature gradient. Equations (8.17) are the most specific constitutive equations for elastic materials (under large deformation and large vibrations in temperature) that can derived from thermodynamic considerations alone. Note that the salient feature of these equations is that, under the most general thermomechanical conditions, the free energy density behaves as a potential function from which the mixed stress tensor $T^{\alpha}{}_i$ and the entropy density of are derivable.

At this point we introduce the alternative approach of deriving the above equations by utilizing directly the results that we derived in Section 3. It was shown previously that the first Law of thermodynamics for an infinitesimal material subsystem could be written in the canonical form

(8.19) $$\dot{\epsilon} = X_i \, \dot{\chi}_i + \dot{Q}$$

where ϵ is the internal energy and θ the heat supply. In eq. (8.19), χ_i have been identified with components of the material stress tensor $\tau^{\alpha\beta}$, through eq.(1.13), and χ_i have been identified with the components of the right Cauchy-Green tensor $C_{\alpha\beta}$, through eq. (1.14). Of course the above interpretation for χ_i and χ_i is not unique ; however, it does enjoy the advantage that equations of state expressed in terms of $\tau^{\alpha\beta}$ and $C_{\alpha\beta}$, remain form-invariant with rigid body motion of the spatial system, provided that q alos remain invariant. This form-invariance is necessary to give the equations of state an objective status, making these completely indifferent to the motion of the observer. This question will be discussed at length in the latter part of this Section. If, however, we choose to consider the constraints imposed by objectively after we deal with the thermodynamic considerations of the problem (as was indeed the case with our exposition of Coleman's approach) then X_n (n = 1, 2, ... , 9) may be identified with $\frac{1}{\rho} T^a_i$ and χ_n (n = 1, 2, . . ,9) may be identified with $y_{i,\,a}$. Then, indeed

(8.20) $$\frac{1}{\rho} T^a_i \, \dot{y}_{i,\,a} = X_n \, \dot{\chi}_n$$

is the rate of work done on the infinitesimal subsystem when this has a unit mass.

But now, eq.'s (3.23a) and (3.23) can be utilized directly to yield eq.'s (8.17a) and (8.17b). Eq. (8.17d) is a direct consequence of the fact that the system is free of thermal constraints as discussed in Appendix I of Section 3. Inequality (8.18) is the result of experimental observation and eq. (8.17c) is a mathematical consequence of ineq. (8.18) following the rationale that if h_i were independent of $\theta,_i$ then evidently values could be given to the former and the latter, such that the inequality (8.18) is violated.

This discussion demonstrates the power of the results derived in Section 3, to the extent that the constitutive equations of elastic materials under large deformation and large variations in temperature can be written down by inspection.

<u>Principle of Objectivity</u>.

Further constraints on the constitutive equations will result as a consequence of the principle of "objectivity", otherwise known as "material indifference" or "isotropy of space". This principle is an axiomatic statement to the effect that the equations of state (8.15) must remain form-invariant with arbitrary time dependent rotation and/or translation of the frame of reference y_i. Thus if y_i is another frame of reference which is obtained from eq. (6.4) and if $T^a{}_i$, ϵ, η, χ, θ and h_i are quantities defined in the new frame y_i, then, as a result of the principle of objectivity eqs. (8.15) must satisfy the constraints :

$$\bar{T}^a_i = T^a_i(\bar{y}_{i,a}, \bar{\theta}) \qquad (8.21a)$$

$$\bar{\epsilon} = \epsilon(\bar{y}_{i,a}, \bar{\theta}) \qquad (8.21b)$$

(8.21c) $$\bar{\eta} = \eta(\bar{y}_{i,a}, \bar{\theta})$$

(8.21d) $$\bar{\psi} = \psi(\bar{y}_{i,a}, \bar{\theta})$$

(8.21e) $$\bar{h}_i = h_i(\bar{y}_{j,a}\ \bar{\theta}, \frac{\partial\bar{\theta}}{\partial\bar{y}_j})$$

Let us consider the effect of the constraints (8.21) on eq. (8.17a). Because $T^a_{\ i}$ is derivable from ψ it is sufficient to consider the constraints (8.21d). Obviously

(8.22) $$\frac{\partial\bar{y}_K}{\partial\chi^a} = Q_{ki}\frac{\partial y_i}{\partial\chi^a} \ .$$

Thus the quantities $\frac{\partial y_i}{\partial\chi^1}, \frac{\partial y_i}{\partial\chi^2}, \frac{\partial y_i}{\partial\chi^3}$ transform as vectors with rotation of the spatial system. Equation (8.21d) requires that ψ, which is a scalar function of the three vectors $\frac{\partial y_i}{\partial\chi^a}$, remain form invariant with rotation of the spatial system.

A fundamental theorem by Cauchy states that this condition will be satisfied if ψ is a function of the scalar products of the three vectors and the determinant $\left|\frac{\partial y_i}{\partial\chi^a}\right|$: Thus

(8.23) $$\psi = \psi\left(\frac{\partial y_i}{\partial\chi^a}\frac{\partial y_i}{\partial\chi^\beta}, \left|\frac{\partial y_i}{\partial\chi^a}\right|, \theta\right)$$

The quantity $y_{ia}y_{i,\beta} \equiv C_{a\beta}$ is known as the right Cauchy-Green tensor. If we require that ψ also remains form invariant with reflection of the system y_i then, since $\frac{\partial y_i}{\partial\chi^a}$ changes sign with reflection, ψ must depend on an even power of $\left|\frac{\partial y_i}{\partial\chi^a}\right|$, say $\left|\frac{\partial y_i}{\partial\chi^a}\right|^2$. However, $\left|\frac{\partial y_i}{\partial\chi^a}\right|^2 = \mathrm{Det}\ C_{a\beta}$ and hence without loss of generality we can say that eq. (8.21d) necessitates that

$$\psi = \psi(C_{\alpha\beta}, \theta). \tag{8.24}$$

It follows immediately that

$$T^{\alpha}_{\ i} = 2\rho \frac{\partial x^{\beta}}{\partial y_i} \frac{\partial \psi}{\partial C_{\alpha\beta}} \tag{8.25}$$

where ψ is a symmetric function (*) of $C_{\alpha\beta}$ and constraint (8.22) is now satisfied. Alternatively,

$$\tau^{\alpha\beta} = 2\rho \frac{\partial \psi}{\partial C_{\alpha\beta}} \tag{8.26}$$

where

$$\tau^{\alpha\beta} = \frac{\partial x^{\alpha}}{\partial y_i} \frac{\partial x^{\beta}}{\partial y_j} T_{ij}. \tag{8.27}$$

i.e., $\tau^{\alpha\beta}$ are the components of the stress tensor in the material system. If ψ is measured per unit undeformed volume then

$$\tau^{\alpha\beta} = \frac{2\rho}{2_o} \frac{\partial \psi}{\partial C_{\alpha\beta}}. \tag{8.28}$$

This result could have been written down directly using eq.'s (1.13), (1.14) and (3.23a).

Note that as a consequence of eq. (8.25) the constraint (8.21a) is

---

(*) See Appendix I to this Section.

automatically satisfied since

$$\bar{T}^a_i = Q_{ij} T^a_j = Q_{ij} 2\rho \frac{\partial \chi^\beta}{\partial y_j} \frac{\partial \psi}{\partial C_{\alpha\beta}} = 2\rho \frac{\partial \chi^\beta}{\partial \bar{y}_i} \frac{\partial \psi}{\partial C_{\alpha\beta}} = T^a_i (\bar{y}_i, a, \theta)$$

(8.29)

The last part of the equation is made possible by the relation

(8.30) $$\bar{C}_{\alpha\beta} = C_{\alpha\beta}$$

as can be easily verified.

On the basis of the above arguments

(8.31) $$\epsilon = \epsilon(C_{\alpha\beta}, \theta)$$

(8.32) $$\eta = \eta(C_{\alpha\beta}, \theta)$$

It may also be shown that constraint (8.21e) is satisfied if

(8.33) $$h_i = \frac{\partial y_i}{\partial \chi^a} h^a(C_{\alpha\beta}, \theta, \theta, a)$$

The form (8.33) follows immediately once one recognizes that the components h^a, with respect to the material system, of the heat flux vector $\vec{h}$, where

(8.34) $$h^a = \frac{\partial \chi^a}{\partial y_i} h_i$$

remain invariant with rotation of the spatial system, and therefore each component can be treated as a scalar function with respect to rigid body rotation of the system y_i. As a result of eqs. (8.18) and (8.21e) and Cauchy's theorem it is easily shown that the following relation must hold

$$h^{\alpha} = h^{\alpha}(C_{\alpha\beta}, \theta, \theta, \alpha) \tag{8.35}$$

from which eq. (8.33) follows immediately.

The heat conduction equation may also be derived readily by using the definition of free energy in conjunction with the First Law and eqs. (8.17a) and (8.17b) and is given by the following eq. (8.36) :

$$h_{i,i} = \rho\theta \left(\frac{\partial \psi}{\partial \theta}\right)^{\cdot} + \rho\dot{q} \tag{8.36}$$

Appendix I.

Proof of equation (8.25). If $\psi = \psi(C_{\alpha\beta})$ it follows as a result of the symmetry of $C_{\alpha\beta}$ that

$$(8.37) \qquad \frac{\partial \psi}{\partial y_{i,\alpha}} = \left(\frac{\partial \psi}{\partial C_{\alpha\beta}} + \frac{\partial \psi}{\partial C_{\beta\alpha}}\right) y_{i,\beta}$$

Let ψ be some function of $C_{\alpha\beta}$, denoted by f, i.e.,

$$(8.38) \qquad \psi = f(C_{\alpha\beta}).$$

Now set

$$(8.39) \qquad f^* = f(C_{\beta\alpha})$$

whereby it follows immediately that

$$(8.40) \qquad \frac{\partial f}{\partial C_{\alpha\beta}} = \frac{\partial f^*}{\partial C_{\beta\alpha}}$$

Substitution of eq. (8.40) in eq. (8.37) yields the result

$$(8.41) \qquad \frac{\partial \psi}{\partial y_{i,\alpha}} = \frac{\partial}{\partial C_{\alpha\beta}} (f + f^*)\, y_{i,\beta}$$

But $f + f^*$ is a symmetric function of $C_{\alpha\beta}$. Thus setting $\psi = \frac{f + f^*}{2}$ one obtains eq. (8.25).

SECTION 9

Constitutive Equations of Dissipative Materials.

We begin with the fundamental laws of thermodynamics, which apply to all continuous media irrespective of their constitutive properties. (For materials that are solid-like, in the sense that they have a memory of their initial configuration, it is more convenient to express these laws in the material coordinate system x^a). In differential form, these are the first law,

$$\dot{\epsilon} = (\rho_o/2\rho)\, \tau^{a\beta}\, \dot{C}_{a\beta} - h^a_{\ ,a} + \dot{q}\,, \tag{9.1}$$

the rate of dissipation inequality,

$$\theta\dot{\gamma} = (\rho_o/2\rho)\, \tau^{a\beta}\, \dot{C}_{a\beta} - \dot{\psi} - \eta\dot{\theta} \geq 0 \tag{9.2}$$

and the heat conduction inequality

$$- h^a\theta_{,a} \geq 0. \tag{9.3}$$

In the case of dissipative materials the stress, the internal energy and entropy densities (and, therefore, the free energy density) of a material neighborhood depend on the entire history of deformation and temperature of that neighborhood.

In the theory of irreversible thermodynamics the effects of history are taken into account by specifying that the stress and free energy density are functions of the current values of $C_{a\beta}$ and θ as well as n additional <u>independent</u> variables q_r, not necessarily observable, called "internal variables". These may be scalars or components of vectors or tensors in the material frame ; whatever their geometric nature they must remain invariant with translation and rotation of the spatial system to satisfy the principle of material indifference. Thus :

(9.4) $$\psi = \psi\,(C_{\alpha\beta}, \theta\,, q_r)$$

(9.5) $$\tau^{\alpha\beta} = \tau^{\alpha\beta}\,(C_{\gamma\delta}\,,\,\theta\,,\,q_r)$$

It has been shown in Section 3 that

(9.6) $$\tau^{\alpha\beta} = \frac{2\rho}{\rho_o}\,\frac{\partial\psi}{\partial C_{\alpha\beta}}$$

(9.7) $$\eta = -\frac{\partial\psi}{\partial\theta}$$

(9.8) $$\theta\dot{\gamma} = -\frac{\partial\psi}{\partial q_r}\,\dot{q}_r \geq 0$$

Furthermore the heat flux vector h^α is a function of $\theta_{,\alpha}$, θ, $C_{\alpha\beta}$ and q_r i.e.,

(9.9) $$h^\alpha = h^\alpha\,(\theta_{,\alpha}, \theta\,,\,C_{\alpha\beta}, q_r)$$

subject to the conditions :

(9.10 a, b) $$h^\alpha \Big|_{\theta_{,\alpha} = 0} = 0,\; h^\alpha\theta_{,\alpha} \leq 0.$$

Finally eq. (9.1) in conjunction with eq.'s (9.6) and (9.7) yields

(9.11) $$h^\alpha{}_{,\alpha} = \theta\left(\frac{\partial\psi}{\partial\theta}\right)^{\boldsymbol{\cdot}} - \frac{\partial\psi}{\partial q_r}\,\dot{q}_r + \dot{q}$$

We note in passing that we could have taken what appears to be a more general view by identifying χ_i with $\rho T^\alpha{}_i$ and χ_i with $y_{i,\alpha}$ (just as we did in Section 8), however, application of the conditions arising out of the axiom of material indifference, would reduce our equations to the form given above.

The remarkable property of the above equations is that they apply to all materials irrespective of their constitution. This has not been generally recognized. In fact the constitutive nature of the material follows from the constitutive equations for q_r, which are called internal constitutive equations.

Thus viscoelastic, plastic and viscoplastic materials differ only in the form of their respective internal constitutive equations for q_α, which link the internal variables to the deformation and temperature histories. We shall proceed to derive constitutive equations for such materials in the following Sections.

SECTION 10.

Constitutive Equations for Viscoelastic Materials.

The internal constitutive equation for these materials was arrived at by the author from purely phenomenological considerations following the rationale that $\frac{\partial \psi}{\partial q_a}$ and $\dot{q}_a$ must be related otherwise values of these quantities could be prescribed independently and in such a fashion that the inequality (9.8) would be violated. Furthermore, since ψ is a function of $C_{\alpha\beta}$, q_r and θ it was concluded that there must exist a set of relations of the type

$$\dot{q}_r = f_r\,(C_{\alpha\beta},\, q_r,\, \theta)\ (r = 1, 2, \ldots, n) \tag{10.1}$$

One may, however, make the form of the above set of relations more specific with the aid of the physical model for the internal variables discussed in Section 7. Since viscoelastic materials are, by and large, polymeric materials which, in turn, are aggregates of long chain molecules, an apt physical counterpart for an internal variable, is a displacement of the junction of two submolecules of a typical molecules of a typical molecule. As a result, eq. (7.33) will apply, in which case eq. (10.1) takes the form

$$\frac{\partial \psi}{\partial q_r} + \eta \dot{q}_r = 0 \tag{10.2}$$

which is much more potent than eq. (10.1) since now the functional form of ψ, the viscocity coefficient η and θ completely determine the stress part of constitutive equation of a viscoelastic material.

Note that, the rate of change of irreversible entropy is now given by the expression

$$\theta\dot{\gamma} = \sum_{r=1}^{n} \eta\dot{q}_r\dot{q}_r \tag{10.3}$$

i.e., γ is a quadratic in the rates of the internal variables.

It would appear that some degree of generality would be gained by writting eq. (10.2) in the form

$$\frac{\partial\psi}{\partial q_r} + b_{rs}\,\dot{q}_s = 0 \tag{10.4}$$

where b_{rs} is a positive definite square matrix. However, the matrix b_{rs} being such, there always exists an orthonormal matrix R_{pq} such that

$$R_{pr}\,R_{qs}\,b_{rs} = \delta_{pq}\,\eta q \text{ (q not summed)} \tag{10.5}$$

One may, therefore define one new set of internal variables q_s such that

$$q_s = R_{st}\,q_t,\ q_t = R_{st}\,q_s \tag{10.6}$$

In terms of q_s (and dropping bars), eq. (10.4) reduces to the form

$$\frac{\partial\psi}{\partial q_r} + \eta_r\,\dot{q}_r = 0 \text{ (r not summed)} \tag{10.7}$$

which is somewhat more general than eq. (10.2). Viscoelastic materials with an internal constitutive equation of the type (10.7) exhibit what one might call internal heterogeneity, since at different points the resistance to motion of a molecule is different

In this case

$$\theta\dot{\gamma} = \sum_{r=1}^{n} \eta_r\,\dot{q}_r\dot{q}_r \tag{10.8}$$

However, we do not wish to tie the theory down to the physical model any more than we have to. Thus, though the model gives q_r a vectorial character we wish to regard eq. (10.7) as the internal constitutive equation of a viscoelastic material, whether q_r are regarded as scalars, vectors or tensors of any order. In fact at this time we are working toward a three dimensional physical model, in which the internal variables appear as second order tensors.

Explicit Constitutive Equations for Viscoelastic Materials under Small Deformation and Small Changes in Temperature.

In what follows we shall consider situations in which the strain in a material region R as well as the temperature change relative to a uniform reference temperature θ_0 are "small". To make the above statement more precise let $\epsilon_{\alpha\beta}(\tau)$ denote the history of the strain tensor (*), for $0 < \tau < t$. Set,

(10.9) $$\|\epsilon_{\alpha\beta}(\tau)\| \equiv \{\epsilon_{\alpha\beta}(\tau)\,\epsilon_{\alpha\beta}(\tau)\}^{1/2}$$

and let the supremum of $\|\epsilon_{\alpha\beta}(\tau)\|$ be Δ.

Similarly let $\vartheta(\tau)$ be the history of the temperature increment relative to the reference temperature θ_0 and let the supremum of $|\vartheta(\tau)|$ be δ. The notion of smallness is made precise by stipulating that $\Delta \ll 1$ and $\delta \ll 1$.

Thus, formally

(10.10) $$\theta = \theta_0 + \vartheta$$

(10.11) $$\eta = \eta_0 + \chi$$

(10.12) $$\|\epsilon_{\alpha\beta}(\tau)\|_{\sup} = \Delta, \quad |\vartheta(\tau)|_{\sup} = \delta$$

(*) $\epsilon_{\alpha\beta} = {}^1/_2\,(C_{\alpha\beta} - \delta_{\alpha\beta})$

To complete the formalism let ψ be the charge in free energy relative to a reference free energy ψ_o, let χ be the entropy change relative to a reference entropy η_o, and let $\sigma_{\alpha\beta}$ denote the stress tensor. The reference state is defined by the condition that $\sigma_{\alpha\beta} = 0$, $\psi = \vartheta = \chi = o$, $q_r = 0$.

Under these conditions, eq.'s (2.4) through (2.9) and eq. (2.10b) become,

$$\hat{\psi} = \psi\,(\epsilon_{\alpha\beta}, \vartheta, q_r) \tag{10.13}$$

$$\sigma_{\alpha\beta} = \frac{\partial \hat{\psi}}{\partial \epsilon_{\alpha\beta}} \tag{10.14}$$

$$\chi = -\frac{\partial \hat{\psi}}{\partial \vartheta} \tag{10.15}$$

$$\theta_o \dot{\gamma} = -\frac{\partial \hat{\psi}}{\partial q_r}\,\dot{q}_r \geq 0 \tag{10.16}$$

$$h_\alpha = k_{\alpha\beta}\,\vartheta,_\beta \tag{10.17}$$

$$h_\alpha\,\vartheta,_\alpha \leq 0 \tag{10.18}$$

where $k_{\alpha\beta}$ is the thermal conductivity tensor.

Finally eq. (2.11), become

$$h_{\alpha,\alpha} = \theta_o\left(\frac{\partial \hat{\psi}}{\partial \vartheta}\right)^{\cdot} - \frac{\partial \hat{\psi}}{\partial q_r}\,\dot{q}_r + \dot{q}. \tag{10.19}$$

It is shown in Appendix I that $|q_r|$ and $|\dot{q}_r|$ may stay small in the sense that given two positive numbers δ_1^r and δ_2^r, however small, Δ and δ can be chosen small enough such that $|q_r| < \delta_1^r$ and $|q_r| < \delta_2^r$.

Explicit constitutive equations for viscoelastic materials under conditions of small strain and small changes in temperature are obtained by expanding ψ in eq. (3.7) in Taylor Series and omitting terms of order higher than $0(\hat{\delta}^2)$(*) ; linear terms must vanish to satisfy the initial conditions.

Before the expansion is carried out, however, it appears desirable to regard q_r not as scalars but components of second order tensors. This, as will be shown, obviates certain difficulties which arise with the representation of fourth order tensors. For instance, in Ref. (15), we were faced with having to assume, without proof, that a fourth order tensor $C_{ijk\ell}$ such that,

(10.20) $$C_{\alpha\beta\gamma\delta} = C_{\beta\alpha\gamma\delta} = C_{\gamma\delta\alpha\beta} = C_{\alpha\beta\delta\gamma}$$

is given by the series

(10.21) $$C_{\alpha\beta\gamma\delta} = \sum_r \frac{a_{\alpha\beta r} a_{\gamma\delta r}}{a_r}$$

where $a_{\alpha\beta r}$ are second order symmetric tensors and a_r are scalars. Problems such as this are obviated by giving the internal variables a tensorial character. Thus the free energy density and other thermodynamic quantities are now functions of $\epsilon_{\alpha\beta}$, and n internal variables $q^r_{\alpha\beta}$ (r = 1,2. . . . n), where $q^r_{\alpha\beta}$ are symmetric second order tensors, with respect to the material system χ_α. In this notation eq.'s(10.19) and (10.7) now read,

(10.22) $$h_{a,a} = \theta_o \left(\frac{\partial \hat{\psi}}{\partial \vartheta}\right) - \frac{\partial \hat{\psi}}{\partial q^r_{\alpha\beta}} \dot{q}^r_{\alpha\beta} + \dot{q}$$

and

(10.23) $$\frac{\partial \hat{\psi}}{\partial q^r_{\alpha\beta}} + \eta^r_{\alpha\beta\gamma\delta} \frac{dq^r_{\alpha\beta}}{dt} = 0 \quad \text{(r not summed)}$$

(*) δ_r is the largest of δ^r_1 and δ^r_2. Also, $\hat{\delta}$ is largest of δ, Δ and δ^r_1.

where $\eta^r_{\alpha\beta\gamma\delta}$ now takes the form of a viscocity tensor which is positive definite for all r, as a result of the ineq. (3.38). It is found experimentally that the viscocity tensor is extremely sensitive to temperature. At this time we shall make the simplifying assumption that all components of all viscocity tensors $\eta^r_{\alpha\beta\gamma\delta}$ (r = 1, 2 . . . . , n) depend on temperature through the same function at $a_T[\vartheta(t)]$ where a_T is known as the "shift factor" in rheological circles. Thus we set

$$\eta^r_{\alpha\beta\gamma\delta} = a_T\, b^r_{\alpha\beta\gamma\delta} \tag{10.24}$$

where $b^r_{\alpha\beta\gamma\delta}$ is independent of strain and temperature for all r. Materials with this property are known as thermorheologically simple.

Equation (10.23) may now be written in the form

$$\frac{\partial\hat{\psi}}{\partial q^r_{\alpha\beta}} + b^r_{\alpha\beta\gamma\delta}\frac{dq^r_{\gamma\delta}}{dz} = 0 \quad \text{(r not summed)} \tag{10.25}$$

where z is given by the expression

$$z = \int_o^t \frac{d\tau}{a_T[\vartheta(\tau)]} \tag{10.26}$$

and is known as the reduced time. Note that

$$\frac{dt}{dz} = a_T \tag{10.27}$$

At this point and in view of my previous discussion, we may write ψ in the form

$$\hat{\psi} = \frac{1}{2}A_{\alpha\beta\gamma\delta}\,\epsilon_{\alpha\beta}\,\epsilon_{\gamma\delta} + \sum_r B^r_{\alpha\beta\gamma\delta}\,\epsilon_{\alpha\beta}\,q^r_{\gamma\delta} + \sum_r C^r_{\alpha\beta\gamma\delta}\,q^r_{\alpha\beta}\,q^r_{\gamma\delta}$$

(*) Expansions of the type $b^{rs}_{\alpha\beta\gamma\delta}\dfrac{dq^r_{\alpha\beta}}{dt}\dfrac{dq^s_{\gamma\delta}}{dt}$ and $A^{rs}_{\alpha\beta\gamma\delta}\,q^r_{\alpha\beta}\,q^s_{\gamma\delta}$ reduced to the above form.

$$+ D_{\alpha\beta}\,\vartheta\epsilon_{\alpha\beta} + \sum_{r} E^{r}_{\alpha\beta}\,\vartheta\, q^{r}_{\alpha\beta} + \frac{1}{2} F \vartheta^2 \tag{10.28}$$

Though, in principle, eq.'s (10.14), (10.15), (10.17), (10.22), (10.25) and (10.28) are sufficient for the derivation of explicit constitutive equations, we shall obtain these only for isotropic materials, so as to keep the algebra at a minimum. For such materials

$$\begin{aligned}
A_{\alpha\beta\gamma\delta} &= A_1\, \delta_{\alpha\beta}\, \delta_{\gamma\delta} + A_2\, \delta_{\alpha\gamma}\, \delta_{\beta\delta} \\
B^{r}_{\alpha\beta\gamma\delta} &= B^{r}_1\, \delta_{\alpha\beta}\, \delta_{\gamma\delta} + B^{r}_2\, \delta_{\alpha\gamma}\, \delta_{\beta\delta} \\
C^{r}_{\alpha\beta\gamma\delta} &= C^{r}_1\, \delta_{\alpha\beta}\, \delta_{\gamma\delta} + C^{r}_2\, \delta_{\alpha\gamma}\, \delta_{\beta\delta} \\
D_{\alpha\beta} &= D\, \delta_{\alpha\beta} \\
E^{a}_{\alpha\beta} &= E^{a}\, \delta_{\alpha\beta} \\
k_{\alpha\beta} &= k\, \delta_{\alpha\beta} \\
b^{r}_{\alpha\beta\gamma\delta} &= b^{r}_1\, \delta_{\alpha\beta}\, \delta_{\gamma\delta} + b^{r}_2\, \delta_{\alpha\gamma}\, \delta_{\beta\delta}
\end{aligned} \tag{10.29 a-f}$$

where b^{r}_2 and $b^{r}_1 + \frac{1}{3} b^{r}_2$ must be positive for all r as a result of ineq. (3.38).

It is worth noting that here we consider materials which are "stable" in the sense that straining of the reference configuration under isothermal conditions will increase the free energy density ψ. Thus $A_{\alpha\beta\gamma\delta}$ and $C^{r}_{\alpha\beta\gamma\delta}$ are positive definite. As a consequence A_o, A_2, C^{r}_o and C^{r}_2 are all positive (*).

(*) In the notation of eq. (10.37).

Omitting superfluous algebra, the coupled thermomechanical constitutive equations take the following form in terms of the hydrostatic stress $\sigma = \frac{\sigma_{aa}}{3}$, the deviatoric stress tensor $s_{a\beta}$, the increment in temperature ϑ, the hydrostatic strain ϵ_{aa}, the deviatoric strain tensor $e_{a\beta}$ and the entropy increment χ; in terms of the above notation :

$$s_{a\beta} = 2 \int_{z_o}^{z} \mu\,(z - z')\frac{\partial e_{a\beta}}{\partial z'}\, dz' \tag{10.30}$$

$$\sigma = \int_{z_o}^{z} K(z - z')\frac{\partial \epsilon_{aa}}{\partial z'}\, dz' + \int_{z_o}^{z} D(z - z')\frac{\partial \vartheta}{\partial z'}\, dz' \tag{10.31}$$

$$-\chi = \frac{\partial \psi}{\partial \vartheta} = \int_{z_o}^{z} D(z - z')\frac{\partial \epsilon_{aa}}{\partial z'}\, dz' + \int_{z_o}^{z} F(z - z')\frac{\partial \vartheta}{\partial z'}\, dz' \tag{10.32}$$

where

$$2\,\mu(z) = (A_2 - \sum_r \frac{B_2^r\, B_2^r}{C_2^r})\, H(z) + \sum_r \frac{B_2^r\, B_2^r}{C_2^r}\, e^{-\rho_r z} \tag{10.33}$$

$$K(z) = (A_0 - \sum_r \frac{B_0^r\, B_0^r}{C_0^r})\, H(z) + \sum_r \frac{B_0^r\, B_0^r}{C_0^r}\, e^{-\lambda_r z} \tag{10.34}$$

$$D(z) = (D - \sum_r \frac{B_r\, E_r}{C_0^r})\, H(z) + \sum_r \frac{B_r\, E_r}{C_0^r}\, e^{-\lambda_r z} \tag{10.35}$$

$$F(z) = (F - \sum_r \frac{E^r\, E^r}{C_0^r})\, H(z) + \sum_r \frac{E^r\, E^r}{C_0^r}\, e^{-\lambda_r z} \tag{10.36}$$

(10.37) $A_0 = 1/3\ (3A_1 + A_2)\ ,\ a_0 = 1/3\ (3a_1 + a_2)\ ,\ C_0^r = 1/3(3C_1^r + C_2^r)$

(10.38) $$\rho_r = \frac{C_2^r}{b_2^r},\ \lambda_r = \frac{C_0^r}{b_0^r}$$

(10.39) $$b_0^r = 3b_1^r + b_2^r$$

Now using eq. (10.19) the heat conduction equation is similarly found to be :

$$\frac{dt}{dz}(k\,\vartheta,_a),_a = \frac{\partial}{\partial z}\int_{z_o}^{z} C_v(z-z')\frac{\partial\vartheta}{\partial z'}\,dz' - \theta_o\frac{\partial}{\partial z}\int_{z_o}^{z} D(z-z')\frac{\partial\epsilon_{aa}}{\partial z'}\,dz'$$

(10.40) $$+ \hat{Q} + \sum_r b_1^r\,\hat{q}_{aa}^r\,\hat{q}_{\beta\beta}^r + \sum_r b_2^r\,\hat{q}_{a\beta}\,\hat{q}_{a\beta}$$

where

(10.41) $$C_v(z) = -\,\theta_o\ F(z)$$

and a roof over a quantity implies differentiation with respect to z and z_o denotes the value of z at the initiation of the deformation. The form of the function $\mu(z)$, K(z) and $C_v(z)$ is, to a large extent, dictated by the thermodynamic inequalities discussed in the foregoing. The fact that b_o^r and C_o^r are all positive means that λ_r are positive ; also since C_2^r and b_2^r are positive, ρ_r must be positive. The above functions consist of a constant term plus a finite some of positive exponential terms which, thus, decay monstonically to zero as $z \to \infty$. In the case of the relaxation function $\mu(z)$ and K(z) one can show that the constant term is also positive as a result of the positive definite nature of ψ. Thus both $\mu(z)$ and K(z) are positive monotonically decreasing functions.

The form of $C_v(z)$ is decided by the thermal stability of the material, which necessitates that the instantaneous specific heat $C_v(o)$ must be positive. Equation (10.36) and (10.41) then show that F must be negative ; it follows that $C_v(z)$ must be a positive monotically increasing function of z, with an asymptotic value of $C_v(\infty) = - F + \sum_r \frac{E^r E^r}{C_o^r}$.

Appendix I.

The form of eq. (10.28) for isotropic materials, according to eq.'s (10.29 a-e), is

$$\hat{\psi} = \frac{1}{2} A_1 \epsilon_{aa}\epsilon_{\beta\beta} + \frac{1}{2} A_2 \epsilon_{a\beta}\epsilon_{a\beta} + B_1^r \epsilon_{aa} q_{\beta\beta}^r + B_2^r \epsilon_{a\beta} q_{a\beta}^r$$

$$+ \frac{1}{2} C_1^r q_{aa}^r q_{\beta\beta}^r + \frac{1}{2} C_2^r q_{a\beta}^r q_{a\beta}^r + D \epsilon_{aa} + E^r q_{aa}^r$$

$$+ \frac{1}{2} F\vartheta^2 \quad \text{(r summed)} \tag{10.42}$$

As a result eq.'s (10.14) and (10.15) yield :

$$\sigma_{a\beta} = A_1 \delta_{a\beta} \epsilon_{aa} + A_2 \epsilon_{a\beta} + B_1^r \delta_{a\beta} q_{aa}^r + B_2^r q_{a\beta}^r + D\vartheta\delta_{a\beta} \quad \text{(r summed)} \tag{10.43}$$

$$-\chi = \frac{\partial\hat{\psi}}{\partial\vartheta} = D\epsilon_{aa} + E^r q_{aa}^r + F\vartheta \quad \text{(r summed)} \tag{10.44}$$

On the other hand,

$$\frac{\partial\hat{\psi}}{\partial q_{a\beta}^r} = B_1^r \epsilon_{aa} \delta_{a\beta} + B_2^r \epsilon_{a\beta} + C_1^r \delta_{a\beta} q_{aa}^r$$

$$+ C_2^r q_{a\beta}^r + E^r \delta_{a\beta} \quad \text{(r not summed)} \tag{10.45}$$

Hence, use of eq. (10.23) in accordance with eq. (10.29f) yields a set of first order differential equations for $q^r_{\alpha\beta}$; these can be expressed as a set for $q^r_{\alpha\alpha}$ and another for the deviatoric part of $q^r_{\alpha\beta}$, which denote by $p^r_{\alpha\beta}$. Thus in the notation of eq.'s (10.37) and (10.39)

$$B^r_0 \, \epsilon_{\alpha\alpha} + C^r_0 \, q^r_{\alpha\alpha} + E^r \vartheta + b^r_0 \frac{dq^r_{\alpha\alpha}}{dz} = 0 \tag{10.46}$$

$$B^r_2 \, e_{\alpha\beta} + C^r_2 \, p^r_{\alpha\beta} + b^r_2 \frac{dp_{\alpha\beta}}{dz} = 0 \tag{10.47}$$

In both eq.'s (10.46) and (10.47) r is not summed. It follows from the above two equations that

$$q^r_{\alpha\alpha} = -\frac{B^r_0}{b^r_0} \int_{z_0}^{z} e^{-\lambda_r(z-z')} \, \epsilon_{\alpha\alpha}(z') \, dz'$$

$$- \frac{E^r}{b^r_0} \int_{z_0}^{z} e^{-\lambda_r(z-z')} \, \vartheta(z') dz' \tag{10.48}$$

$$p^r_{\alpha\beta} = -\frac{B^r_2}{b^r_2} \int_{z_0}^{z} e^{-\rho_r(z-z')} e_{\alpha\beta}(z') dz' \tag{10.49}$$

where λ_r and ρ_r are given by eq. (10.38).

In the light of the tensorial notation that we have adopted for the internal variables, let

$$\|\epsilon_{\alpha\beta}\|_{\sup} = \Delta \, , \; \|e_{\alpha\beta}\|_{\sup} = \Delta_1 \; , \; |e_{\alpha\alpha}|_{\sup} = \Delta_0 \tag{10.50}$$

where, $\|\epsilon_{\alpha\beta}\| = |\epsilon_{\alpha\beta} \, \epsilon_{\alpha\beta}|^{1/2}$, etc. Evidently,

$$\Delta^2 = \Delta_1^2 + \frac{1}{3} \Delta_0^2 \tag{10.51}$$

Then as a result of eq.'s (10.49) and (10.50)

(10.52) $$||P_{ij}|| \leq \frac{|B_2^a|}{C_2^a} \Delta_1$$

(10.53) $$|q_{aa}^a| \leq \frac{|B_0^r|}{C_0^r} \Delta_0 + \frac{|E^r|}{C_0^r} \delta$$

where as before $| .\vartheta |_{sup} = \delta$

Also from eq. (10.47),

(10.54) $$(b_2^r)^2 \left\| \frac{dp_{\alpha\beta}}{dz} \right\|^2 = (B_r^2)^2 \, ||e_{\alpha\beta}||^2 + (C_r^2)^2 \, ||P_{\alpha\beta}||^2 + 2B_r^2 \, C_r^2 \, |P_{\alpha\beta} \, e_{\alpha\beta}|$$

However, since

(10.55) $$\left|P_{\alpha\beta} \, e_{\alpha\beta}\right| \leq \left\|P_{\alpha\beta}\right\| \, \left\|e_{\alpha\beta}\right\|$$

it follows from (10.54) that

(10.56) $$b_2^r \left\| \frac{dp_{\alpha\beta}}{dz} \right\| \leq 2\,\Delta_1 \left|B_2^r\right|$$

Also as a result of eq. (10.46)

(10.57) $$b_0^r \left| \frac{d\overset{r}{q}_{kk}}{dz} \right| \leq 2 \left|B_0^r\right| \Delta_0 + 2\left|E^r\right| \delta$$

At this point we order our internal variables as shown,

$$\overset{1}{p}_{\alpha\beta} \, , \overset{2}{p}_{\alpha\beta} \cdots\cdots \overset{m}{p}_{\alpha\beta} \, ; \overset{1}{q}_{aa} \, , \overset{2}{q}_{aa} \cdots\cdots \overset{m}{q}_{aa} \, .$$

Let q_r be a typical internal variable. Then, whether it belongs to the p-group or the q-group above, as a result of eq.'s (10.52), (10.53), (10.56) and (10.57), given two positive member δ_1 and δ_2, however small, we can choose Δ_o and Δ_1 (and therefore Δ) and δ such that

$$\left|q_r\right| \leq \delta_1 \quad \text{and} \quad \left|\frac{dq_r}{dz}\right| \leq \delta_2 \,. \tag{10.58a,b}$$

Viscoelastic Materials under Large Iosthermal Deformation.

Because the constitutive equations of viscoelastic materials under large deformation have a rather complex analytical representation, especially if the thermal conditions are not steady, we shall restrict ourselves to the problem of isothermal deformation. If, in addition, the internal variables are given a tensorial character, the form of the constitutive equations become (following eq.'s (9.6), (9.7) and (10.2))

(10.59)
$$\tau^{\alpha\beta} = \frac{2\rho}{\rho_0} \frac{\partial \psi}{\partial C_{\alpha\beta}}$$

(10.60)
$$\eta = - \left(\frac{\partial \psi}{\partial \theta}\right)_{\theta=\theta_0}$$

(10.61)
$$\frac{\partial \psi}{\partial q^{(r)}_{\alpha\beta}} + \eta \; \dot{q}^{(r)}_{\alpha\beta} = 0$$

where θ_0 is the reference equilibrium temperature. Equation (10.61) was arrived at using, eq. (10.2) rather than (10.23) ; furthermore the form of eq. (10.2) was retained for large deformation as well. Some comments are in order regarding the above equations. Choosing eq. (10.2) has the physical implication that resistance to motion of points interior to an infinitesimal element, in proportional to the velocity of the point for large deformation as well. This is not unreasonable, because it is possible for the deformation to be large and yet sufficiently slow to warrant the above assumption. The choice of a single scalar viscocity coefficient η rather than a viscocity tensor $\eta_{\alpha\beta\gamma\delta}$ implies that the viscous resistance to motion in shear involves the same η as that in dilatation. Finally the choice of a single viscocity coefficient for all $q^{(r)}_{\alpha\beta}$ implies a homogeneous viscous resistance to motion inside an infinitesimal material element.

Under the present conditions of large deformation explicit forms of ψ are not easily arrived at. We know that the strains $\frac{1}{2}(C_{\alpha\beta} - \delta_{\alpha\beta})$ are not small and, at this point, we cannot say under what conditions, if any, $q^{(r)}_{\alpha\beta}$ and/or $\dot{q}^{(r)}_{\alpha\beta}$ will be small. Therefore, we have to proceed heuristically.

Two obvious choices present themselves. The first is to assume that $q^{(r)}$ and $\dot{q}^{(r)}$ are small, expand ψ again in a Taylor series in $q^{(r)}$ but make use of this information in the expansion, (thus terminating the latter at the quadratic term) and then examine the physical situations appropriate to the above assumption. The second is to expand ψ formally, in a Taylor series in C and $q^{(r)}$ (assuming requisite smoothness), where the tilde under a symbol denotes a tensor, and to proceed from there. We proceed with the first choice, and set

$$\psi = \psi_o(\underset{\sim}{C}) + \psi_r^{\alpha\beta}(\underset{\sim}{C})\, q^{(r)}_{\alpha\beta} + \frac{1}{2}\psi_{rs}^{\alpha\beta\gamma\delta}\, q^{(r)}_{\alpha\beta}\, q^{(s)}_{\beta\gamma} \qquad (10.62)$$

where explicit dependence of the coefficients on θ_o has been suppressed. Since q is symmetric ψ_{rs} is symmetric in the indices α, β as well as γ, δ ; it is also symmetric with respect to the pairs of indices $(\alpha\beta)$ and $(\gamma\beta)$.

Just as before we shall regard the material as stable in the sense that positive work on an infinitesimal element must be done to change its configuration. Thus, of necessity, the coefficient $\psi_{rs}^{\alpha\beta\gamma\delta}$ must be a positive definite matrix for fixed values of the upper indices and a positive definite fourth order tensor for fixed values of the lower indices ; furthermore we assume that this coefficient is independent of C.

Letting

$$\pi^{\alpha\beta} = \frac{\rho_0}{2\rho}\,\tau^{\alpha\beta} \qquad (10.63)$$

and as a result of eq.'s (10.5(0, (10.61) and (10.62) we obtain,

(10.64)
$$\pi^{\alpha\beta} = \frac{\partial\psi_0}{\partial C_{\alpha\beta}} + \overset{(r)}{q}_{\mu\nu} \frac{\partial\psi_r^{\mu\nu}}{\partial C_{\alpha\beta}}$$

(10.65)
$$\eta\dot{\overset{(r)}{q}}_{\alpha\beta} + \psi_{rs}^{\alpha\beta\gamma\delta}\,\overset{(s)}{q}_{\gamma\delta} + \psi_r^{\alpha\beta}(\underset{\sim}{C}) = 0$$

Note that ψ_{rs} may be diagonalized by means of an appropriate transformation. Thus, without loss of generality equation (10.65) may be written in the form :

(10.66)
$$\eta\dot{\overset{(r)}{q}}_{\alpha\beta} + \psi_r^{\alpha\beta\gamma\delta}\,\overset{(r)}{q}_{\gamma\delta} + \psi_r^{\alpha\beta} = 0$$
(r not summed)

We shall limit the discussion to isotropic materials. In this case

(10.67)
$$\psi_r^{\alpha\beta\gamma\delta} = \frac{1}{3}\overset{(r)}{\rho_1}\delta_{\alpha\beta}\,\delta_{\gamma\delta} + \overset{(r)}{\rho_2}\delta_{\alpha\gamma}\,\delta_{\beta\delta}$$

and, therefore, eq. (10.49) becomes :

(10.68)
$$\eta\dot{\overset{(r)}{q}}_{\alpha\beta} + \frac{1}{3}\overset{(r)}{\rho_1}\,\overset{(r)}{q}_{\mu\mu}\,\delta_{\alpha\beta} + \overset{(r)}{\rho_2}\,\overset{(r)}{q}_{\alpha\beta} + \psi_r^{\alpha\beta} = 0$$
(r not summed)

The left hand side of eq. (10.68) may be written in terms of its deviatoric and hydrostatic parts, accordingly, we let $\underset{\sim}{\hat{q}}$ and q denote respectively the deviatoric and hydrostatic part of q and similarly we let ϕ_r and ϕ_r have the same connotation. Then eq. (10.68) yields :

(10.69a)
$$\eta\dot{q}^{(r)} + (\rho_1 + \rho_2)^{(r)}\,q^{(r)} + \phi^{(r)} = 0$$

(10.69b)
$$\eta\dot{\underset{\sim}{\hat{q}}}^{(r)} + \rho_2^{(r)}\,\underset{\sim}{\hat{q}}^{(r)} + \underset{\sim}{\phi}^{(r)} = 0$$

Integration of eq.'s (10.69a) and (10.69b) yields the following constitutive equation

$$\pi^{\alpha\beta} = \frac{\partial\phi_0}{\partial C_{\alpha\beta}} + \sum_{r=1}^{n} \frac{1}{\rho_2^{(r)}} \frac{\partial\phi_{\mu\nu}^{(r)}}{\partial C_{\alpha\beta}} \int_0^t e^{-\overset{r}{a_2}(t-\tau)} \frac{\partial\phi_{\mu\nu}^{(r)}}{\partial\tau}\,d\tau$$

$$+ \sum_{r=1}^{n} \frac{1/3}{\rho_1^{(r)}+\rho_2^{(r)}} \frac{\partial \phi^{(r)}}{\partial C_{\alpha\beta}} \int_o^t e^{-a_1^r(t-\tau)} \frac{\partial \phi^{(r)}}{\partial \tau} d\tau \tag{10.70}$$

where

$$\phi_o = \psi_o - \sum_{r=1}^{n} \frac{1}{2\rho_2^{(r)}} \phi_{\mu\nu}^{(r)} \phi_{\mu\nu}^{(r)} + \frac{1}{6(\rho_1^{(r)}+\rho_2^{(r)})} \phi^{(r)} \phi^{(r)}, \tag{10.71}$$

$$\phi_{\mu\nu}^{(r)} = \psi_{\mu\nu}^{(r)} - \frac{1}{3} \delta_{\mu\nu} \psi_{\lambda\lambda}^{(r)}, \ \phi^{(r)} = \psi_{\mu\mu}^{(r)} \tag{10.72a,b}$$

$$a_1^r = (\rho_1 + \rho_2)^r/\eta, \ a_2 = \rho_2^{(r)}/\eta \tag{10.73a,b}$$

Before we engage in a lengthy discussion of eq. (10.70) we point out some of its important properties. For instance when the material has reached an equilibrium state following some deformation history, the hereditary part of eq. (10.70) vanishes and

$$\pi^{\alpha\beta} = \frac{\partial \phi_o}{\partial C_{\alpha\beta}} \tag{10.74}$$

On the other hand at small times following a rapid deformation history $q_{rs}^{(r)} \sim 0$ and

$$\pi^{\alpha\beta} = \frac{\partial \psi_o}{\partial C_{\alpha\beta}} \tag{10.75}$$

In other wards, ϕ_o and ψ_o are equilibrium and instantaneous potentials, respectively. The existence of these potentials has also been shown by Coleman and Lianis.

However, one can easily show, that a potential (albeit time dependent) also exists under conditions of relaxation. This result does not follow from other theories such as Coleman's, though is used as a priori assumption in the BKZ theory. Under relaxation conditions,

$$\phi^{(r)}_{\mu\nu} = \phi^{(r)}_{\mu\nu o}\ H(t),\ \phi^{(r)} = \phi^{(r)}_{o}\ H(t). \tag{10.76}$$

Equation (10.76) in conjunction with eq. (10.70) indeed yields the following result under conditions of relaxation :

$$\pi^{\alpha\beta} = \frac{\partial W(C_{\alpha\beta}, t)}{\partial C_{\alpha\beta}} \tag{10.77}$$

where

$$W = \phi_o + \frac{1}{2}\Sigma_r \{ \frac{1}{\rho_2^{(r)}} \phi^{(r)}_{\mu\nu o}\, \phi^{(r)}_{\mu\nu o}\, e^{-a_2^r(t)} + \frac{1/3}{\rho_2^{(r)}+\rho_2^{(r)}}\, \phi^{(r)}_{o}\, \phi^{(r)}_{o}\, e^{-a_1^r t} \} \tag{10.78}$$

Equation (10.77) has had strong experimental corroboration by T. Smith[21], R. F. Landel[22], Thirion[23] and others.

The constitutive equation (10.70) is too general to be useful under more complex histories of deformation. We shall seek therefore, physically meaningful ways to simplify it. We shall offer two simplified versions of eq. (10.70).

The first follows from the experimental observation that under conditions of relaxation and for moderate strain $W(C_{\alpha\beta}, t)$ may be expressed in the factored form

$$W(C_{\alpha\beta}, t) = W_o(C_{\alpha\beta})\ f(t) + g(t) \tag{10.79}$$

where g(t) is an immaterial function of time.

There are various ways of reducing (10.78) to (10.79). However, it appears essential that $\psi^{(r)}_{\mu\nu}$ should be of the form

$$\psi^{(r)}_{\mu\nu} = \psi_{\mu\nu} a_r\ ;\ \phi^{(r)}_{\mu\nu} = \phi^{(\dot{\tilde{C}})}_{\mu\nu} a_r\ ,\ \phi^{(r)} = a_r \phi \tag{10.80}$$

where a_r are independent of the deformation tensor $C_{\alpha\beta}$.

Furthermore, either ρ_1^r must equal ρ_2^r or ϕ must be a function only of I_3, and the material must be incompressible in which case the term

$$3/2 \sum_{r=1}^{n} \phi^2 \frac{e^{-a_1{}^r t}}{\rho_1{}^{(r)}+\rho_2{}^{(r)}}$$

becomes a non-essential part of W, as far as $\pi^{\alpha\beta}$ is concerned.

The constraint (10.80) yields the equation :

$$\pi^{\alpha\beta} = \frac{\partial\phi_0}{\partial C_{\alpha\beta}} + \frac{\partial\phi_{\mu\nu}}{\partial C_{\alpha\beta}} \sum_{r=1}^{n} \frac{a_r{}^2}{\rho_2{}^{(r)}} \int_0^t e^{-a_2{}^{(r)}(t-\tau)} \frac{\partial\phi_{\mu\nu}}{\partial\tau} d\tau$$

$$+ \frac{\partial\phi}{\partial C_{\alpha\beta}} \sum_{r=1}^{n} \frac{1/3 a_r{}^2}{\rho_1{}^{(r)}+\rho_2{}^{(r)}} \int_0^t e^{-a_1{}^{(r)}(t-\tau)} \frac{\partial\phi}{\partial\tau} d\tau \tag{10.81}$$

In the limit of small deformation, and in view of eq. (10.45)

$$\psi_{\mu\nu} = E_{\mu\nu} = \frac{1}{2}(C_{\mu\nu} - \delta_{\mu\nu}), \ \phi = E_{\alpha\alpha} \tag{10.82}$$

where $E_{\mu\nu}$ is the Green deformation tensor. As a result of eq. (10.82),

$$\sum_{r=1}^{n} \frac{a_r{}^2}{\rho_2{}^{(r)}} e^{-a_2{}^{(r)}t} = G(t) - G_\infty \tag{10.83}$$

$$\sum_{r=1}^{n} \frac{1/3 a_r{}^2}{\rho_1{}^{(r)}+\rho_2{}^{(r)}} e^{-a_1{}^{(r)}t} = K(t) - K_\infty \tag{10.84}$$

where G(t) and K(t) are the shear and bulk moduli of linear viscoelasticity and G_∞ and K_∞ are their equilibrium values. Substitution of eq.'s (10.83) and (10.84) in eq. (10.81) finally yield the following constitutive equation :

$$\pi^{\alpha\beta} = \frac{\partial\psi_\infty}{\partial C_{\alpha\beta}} + \frac{\partial\phi_{\mu\nu}}{\partial C_{\alpha\beta}} \int_0^t G(t-\tau) \frac{\partial\phi_{\mu\nu}}{\partial\tau} d\tau + \frac{\partial\phi}{\partial C_{\alpha\beta}} \int_0^t K(t-\tau) \frac{\partial\phi}{\partial\tau} d\tau \tag{10.85}$$

where ψ_∞ is the form of the free energy at equilibrium.

As a result of the assumption of material isotropic ψ_∞, ϕ and $\phi_{\mu\nu}$ are isotropic scalar and tensor functions respectively of $C_{\alpha\beta}$. Explicitly, the most general forms of these functions are :

(10.86) $$\phi_{\mu\nu} = f_0\ \delta_{\mu\nu} + f_1\ C_{\mu\nu} + f_2\ C_{\mu\lambda}\ C_{\lambda\mu}$$

(such that $\phi = 0$)

(10.87) $$\phi = \phi(C_1, C_2, C_3)$$

(10.88) $$\psi_\infty = \psi_\infty(C_1, C_2, C_3)$$

where f_1 are functions of C_1, C_2, C_3, the principal invariants of C, where

(10.89a) $$C_1 = C_{\alpha\alpha}$$

(10.89b) $$C_2 = \frac{1}{2}(C_{\alpha\alpha}\ C_{\beta\beta} - C_{\alpha\beta}\ C_{\alpha\beta})$$

(10.89c) $$C_3 = \mathrm{Det}\ (C_{\alpha\beta})$$

If the material is incompressible and in view of our previous discussion, $\phi =$ $= \phi(C_3)$ (*), in which case eq. (10.85) becomes :

(10.90) $$\tau^{\alpha\beta} = 2\frac{\partial\psi_\infty}{\partial C_{\alpha\beta}} + 2\frac{\partial\phi_{\mu\nu}}{\partial C_{\alpha\beta}}\int_0^t G(t-\tau)\frac{\partial\phi_{\mu\nu}}{\partial\tau}d\tau + pC^{\alpha\beta}$$

where p is an indeterminate hydrostatic pressure $C^{\alpha\beta}$ is the inverse of tensor $C_{\alpha\beta}$ and $C_3 = 1$. But now we can recognize immediately that

(10.91) $$\phi_{\mu\nu}[\underset{\sim}{C}(t)]\ \phi_{\mu\nu}[\underset{\sim}{C}(\tau)] \equiv \Omega\,(\underset{\sim}{C}(t), \underset{\sim}{C}(\tau))$$

(*) Otherwise the limit $K(t-\tau)\frac{\partial\phi}{\partial\tau}$, $K\to\infty, \frac{\partial C_3}{\partial\tau}\to 0$ (for all τ) does not exist.

where Ω is a scalar symmetric function of C(t) and C(τ) ; for isotropic materials, in fact, Ω can only be a function of the joint invariants of C(t) and C(τ).

Then eq. (10.90) assumes the elegant form

$$\tau^{\alpha\beta} = \frac{\partial}{\partial C_{\alpha\beta}} \quad \psi_\infty + \int_0^t G(t-\tau) \frac{\partial \Omega}{\partial \tau} d\tau + pC^{\alpha\beta} \qquad (10.92)$$

For materials obeying eq. (10.92) the stress (whithin an arbitrary hydrostatic pressure) is always derivable from a history dependent potential.

For instance, this was found (24) to be the case for dimethyl siloxane rubber containing 28 wt.–$\%_o S_i 0_2$ as filler, under conditions of room temperature.

The experimental data involving uniaxial and strip biaxial tests, were described extremely well by the functions

$$\psi_\infty = 0,$$
$$\Omega = k_1\ \partial_1(t)\ \partial_1(\tau) - k_2\ \{\partial_2(t) + \partial_2(t)\} + 2k_3\ \{\partial_1(t) + \partial_1(\tau)\} \qquad (10.93)$$

where $\partial_1 = trE$ and $\partial_2 = trE^2$, and K_1, K_2 and K_3 are nondimensional constants. This is shown in Fig. 1 and 2 at the end of this section.

In fact it was found that in the time range of measurements which was between 0.5 and 30 minutes :

$$G(t) = 63.6 \quad \frac{t}{30}^{-.03} \quad g/mm^2 \qquad (10.94)$$

$$k_1 = 1.47, k_2 = 1.40, \ k_3 = 1.22 \qquad (10.95)$$

For compressible materials, eq. (10.85) applies in which case Ω is composed of three potentials, Ω_1 and Ω_2 and ψ_∞ where

$$\Omega_1 = \phi_{\mu\nu}(t)\, \phi_{\mu\nu}(\tau),\ \Omega_2 = \phi\,(t)\, \phi\,(\tau) \tag{10.96a,b}$$

In this event

$$\pi^{\alpha\beta} = \frac{\partial}{\partial C_{\alpha\beta}} \quad \psi_\infty(\underset{\sim}{C}) + \int_0^t G(t-\tau)\frac{\partial \Omega_1}{\partial \tau}\,(\underset{\sim}{C},\, \underset{\sim}{C}(\tau))\,d\tau$$

$$+ \int_0^t K(t-\tau)\frac{\partial}{\partial \tau}\Omega_2\,(\underset{\sim}{C},\, C(\tau))\,d\tau \tag{10.97}$$

The Coleman Constitutive Equation.

This equation can be arrived at in an approximate sense from eq. (10.53) by assuming that the deformation history has been continuous as well as slow in the past. If the first condition is met then

$$\frac{\partial \overset{(r)}{\phi}_{\mu\nu}}{\partial \tau} = \frac{\partial \overset{r}{\phi}_{\mu\nu}}{\partial C_{\gamma\delta}}\,\frac{\partial C_{\gamma\delta}}{\partial \tau} \tag{10.98}$$

and the first factor on the right hand side of eq. (10.98) may be taken (approx.) outside the integral. In this case eq. (10.70) reads :

$$\pi^{\alpha\beta} = \frac{\partial \phi_0}{\partial C_{\alpha\beta}} + \sum_{r=1}^{n} \frac{1}{\rho_2{}^{(r)}}\,\frac{\partial \overset{(r)}{\phi}_{\mu\nu}}{\partial C_{\alpha\beta}}\,\frac{\partial \overset{r}{\phi}_{\mu\nu}}{\partial C_{\gamma\delta}} \int_0^t e^{-a_2{}^{(r)}(t-\tau)}\,\frac{\partial C_{\gamma\delta}}{\partial \tau}\,d\tau$$

$$+ \sum_{r=1}^{n} \frac{1/3}{(\rho_1+\rho_2)^{(r)}}\,\frac{\partial \phi^{(r)}}{\partial C_{\alpha\beta}}\,\frac{\partial \phi^{(r)}}{\partial C_{\gamma\delta}} \int_0^t e^{-a_1{}^{(r)}(t-\tau)}\,\frac{\partial C_{\gamma\delta}}{\partial \tau}\,d\tau \tag{10.99}$$

If one recalls that $\phi^{(r)}_{\mu\nu}$ are isotropic functions of $C_{\alpha\beta}$ and that $\phi^{(r)}$ are functions of the invariants of C, then, after some algebra, one obtains the following result in matrix notation

$$\pi = h_1 I + 2h_2 C + 3h_3 C^2 + 2\int_{-\infty}^{t} a_{01}(t-\tau)\frac{\partial C}{\partial \tau}d\tau$$

$$+ C\int_{-\infty}^{t} a_{02}(t-\tau)\frac{\partial C}{\partial \tau}d\tau + \int_{-\infty}^{t} a_{02}(t-\tau)\frac{\partial C}{\partial \tau}d\tau C$$

$$+ C^2\int_{-\infty}^{t} a\ (t-\tau)\frac{\partial C}{\partial \tau}d\tau + \int_{-\infty}^{t} a_{03}(t-\tau)\frac{\partial C}{\partial \tau}d\tau C^2 \tag{10.100}$$

$$+ [I C C^2]\int_{-\infty}^{t} [a_{ij}(t-\tau)] \begin{bmatrix} \mathrm{tr}\ \partial C/\partial\tau \\ \mathrm{tr}\ \{C(t)\partial C/\partial\tau\} \\ \mathrm{tr}\ \{C^2(t)\partial C/\partial\tau\} \end{bmatrix} d\tau$$

where $h_a = \dfrac{\partial\phi_o}{\partial J_A}$ and a_{ij} are functions of the invariants J_A which are given by eq. (10.84) i.e. ;

$$J_1 = \mathrm{tr}C,\ J_2 = \mathrm{tr}C^2,\ J_3 = \mathrm{tr}C^3\ , \tag{10.101}$$

which was given by Coleman (25) and Lianis (26) who showed from the differentiability of functionals and the thermodynamic results of Coleman that $a_{ij} = a_{ji}$. This last result follows immediately from eq. (10.82).

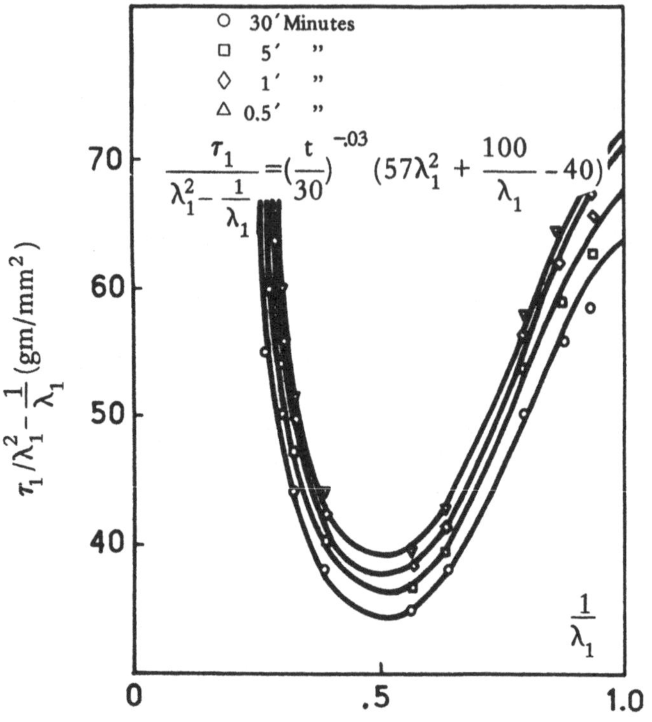

Fig. 1. Uniaxial stress versus inverse axial extension ratio.

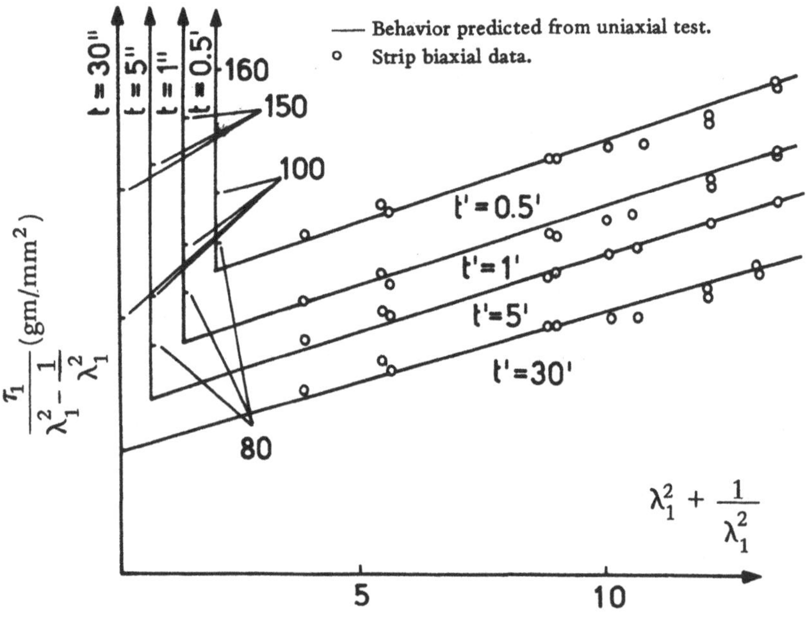

Fig. 2. Strip biaxial behavior. Load applied in direction 1.

REFERENCES

1 Meixner, J., IUTAM Symposium on "Irreversible Aspects of Continuum Mechanics", Vienna, 1966.

2 Perzyna, P., Plochoski, Z. and Valanis, K. C., "Thermomechanical Problems of Continuous Media", Polish Academy of Science, Warsaw, 1970.

3 Meixner, J., Ann der Physik 5, 244, 1943.

4 Biot, M.A., Jour. App. Phys. 25, 1385, 1954.

5 Eringen, A. C., Phys. Rev. 117, 1174,1961.

6 Schapery, R.A., J. App. Phys. 35, 1451, 1964.

7 Kluitenberg, G. A., Physica, 30, 1945, 1964.

8 Coleman, B.D., Arch. Rat. Mech. Anal. 17, 1, 1964.

9 Truesdell, C. and Toupin, R., Handbuch der Physik, III/1, Springer-Verlag, 1960.

10 Echart, C., Phys. Rev. 58, 267, 1940.

11 Volterra, V., Theory of Functionals, Dover, N. Y. 1958.

12 Caratheodory, C., Math. Annal., 67, 355, 1909.

13 Valanis, K.C., Jour. Math. and Phys. 45, 197, 1966.

14 Valanis, K.C., Iowa State Un. Bull, Rep. 52, Jan. 1967.

15 Valanis, K.C , Jour. Math. and Phys. 46, 164, 1967.

16 Valanis, K.C., Mechanical Behavior Under Dynamic Loads, Springer-Verlag, N. Y., 1968.

17 Valanis, K.C., Jour. Math and Phys. 47, 262, 1968.

18 Valanis, K.C., "Irreversibility and Existence of Entropy", Journ. Non-lin. Mechanics, (in press).

19 Ferry, J.D., Viscoelastic Properties of Polymers, Wiley, N. Y., 1970.

20 Rouse, P.E., J. Chem Phys. 21, 1272, 1953.

21 Smith, T.L. and Frederick, J.E., J. App. Phys. 36, 2996, 1965.

22 Landel, R.F. and Stedry, P.J., J. App. Phys. 31, 1885, 1960.

23 Chasset, R. and Thirion, P., Proc. Intern. Conf. Phys. Non-Crystal. Solids, Amsterdam, 1965.

24 Valanis, K.C. and Landel, R.F., Trans. Soc. Rheol. 11, 243, 1967.

25 Coleman, B.D. and Noll, W., Rev. Mod. Phys., 33, 239, 1961.

26 Lianis, G., Rec. Adv. Eng. Sci., 3, 499, 1968.

ADDENDUM

A Theory of Viscoplasticity without a Yield Surface

Very recently we have extended the application of the theory of internal variables to the constitutive equations of viscoplastic materials. Because of lack of time we were not able to incorporate this work in the body of the lectures. As a consequence we present it as a self-contained addendum, to underline the broad applicability of the internal variables. The notation is somewhat different from that of the rest of the lectures, but is explained in the text.

The addendum is presented in two parts. In Part I, we propose a mathematical theory of thermo-viscoplasticity which is a synthesis of experimentally observed material behavior on one hand, and the concepts of irreversible thermodynamics on the other. The underlying principle is that the history of deformation is defined in terms of a "time scale" which is not measured by a clock, but is in itself a property of the material at hand.

The theory is unifying in the sense that theories of plasticity, viscoelasticity and elasticity can be obtained from it as special cases.

In Part II we use the theory of Part I, to give quantitative analytical predictions on the mechanical response of aluminum and copper under conditions of complex strain histories. One single constitutive equation describes with remarkable accuracy and ease of calculation diverse phenomena, such as cross-hardening, loading and unloading loops, cyclic hardening as well as behavior in tension in the presence of a shearing stress, which have been observed experimentally by four different authors.

PART I

GENERAL THEORY

SECTION 1

Endochronic Theory of Viscoplasticity

In current theories of plasticity, to explain the observed discontinuities in material behavior upon loading beyond the "yield point" and upon unloading, one has to introduce the concept of a yield surface in stress space as well as a "loading function" to distinguish between loading and unloading. Similarly, in the case of viscoplasticity, the existence of a static stress-strain relation and a yield surface are assumed and the stress increment, with respect to the static value, is related to the strain rate, or more generally to the strain history, by a constitutive equation.

However, the fact that the phenomenon of yield is usually a gradual transition from a linear to a non-linear stress-strain response, makes it difficult to say precisely where yield has occurred, to the extent that different definitions of yield are used for this purpose. Three such definitions, for instance, are (a) the deviation from linearity in the relation between some measure of strain and stress (b) the intersection of the initial part of a stress strain curve and the backward linear extrapolation of the "plastic" part of the curve and (c) a value of "proof" stress corresponding to an arbitrarily defined value of "proof" strain.

Though, from an engineering viewpoint, the initial yield surface is not overly influenced by the definition of yield, it has been found experimentally that subsequent yield surfaces of a strain hardening material are influenced by the definition of yield to an extraordinary degree. (See Appendix I). If we insist that the increment in plastic strain is to be normal to the yield surface, then, for complex stress histories, each such definition will give rise to a different plastic strain history. Only one of these can be the correct one.

The conceptual difficulties that are encountered by the introduction of the yield surface are completely circumvented by our theory of plasticity which is developed on the basis of the observation that the state of stress in the neighborhood of a point in a plastic material depends on the set of all previous states of deformation of that neighborhood, but it does not depend on the rapidity at which such deformation states have succeeded one another (*).

The independence of stress of the rapidity of succession of deformation states is achieved by introducing a time scale ξ which is independent of t, the external time measured by a clock, but which is intrinsically related to the deformation of the material.

Of course there are many ways of introducing such a time scale. However, it appears almost mandatory that ξ should be a monotonically increasing function of deformation, otherwise two different states of deformation could exist "simultaneously" i.e. for the same value of ξ. Furthermore, a positive rate of change $\frac{d\xi}{d\epsilon}$, of the internal energy density ϵ with respect to ξ could not be interpreted unambiguously as a process of increasing ϵ, if $d\xi$ could be negative.

A logical definition (**) for ξ is then given by the relation

$$d\xi^2 = P_{ijk\ell} dC_{ij} dC_{k\ell} \tag{3.1}$$

(*) In the present Section and in subsequent Sections (with the exception of Section 2) we shall assume that mechanical changes take place in constant temperature environment, such as an isothermal atmosphere. The thermal changes in the material will, therefore, be mechanically induced and, in general, will remain small, Conversely, only thermal changes of this nature will be considered in this paper.

(**)Alternative but less general definitions have appeared in the literature. For instance, Ilyushin (1) and later Rivlin (2) defined a "time" s by the relation $ds^2 = dC_{ij} dC_{ij}$. However, we have found that this definition is too narrow to describe, quantitatively, material behavior in the plastic range as will be discussed later. The effect of temperature on ξ, will at this time, be considered sufficiently small to be negligible. For a more vague allusion to this possibility see also, Schapery (3).

where C_{ij} is the Right Cauchy-Green tensor and $P_{ijk\ell}$ is a fourth order tensor which could depend on C_{ij}. The positive definite nature of $d\xi^2$ requires that $P_{ijk\ell}$ be positive definite. In the case of small deformation.

$$d\xi^2 = P_{ijk\ell} d\epsilon_{ij} d\epsilon_{k\ell}$$

where ϵ_{ij} is the small deformation strain tensor and $P_{ijk\ell}$ could depend on ϵ_{ij}.

Actual materials, on the other hand, do, in general, depend on the history of deformation as well as on the rapidity, or rate, at which deformation states succeed one another. To describe materials of this type one may construct a theory of viscoplasticity by introducing a time scale ζ which is related to the external time t.

It appears logical to define ζ by the relationship

$$d\zeta^2 = \alpha^2 d\xi^2 + \beta^2 dt^2$$

where α and β are scalar material parameters. Henceforth $d\zeta$ will be called an "intrinsic time measure", and $z(\zeta)$, such that $\frac{dz}{d\zeta} > 0$ $(0 < \zeta < \infty)$, will be called an "intrinsic time scale".

In our theory, the stress (among other properties) is necessarily, a functional of the strain history, defined with respect to the intrinsic time scale, the latter being a property of the material at hand. As a result we have called our theory an endochronic theory of viscoplasticity.

The theory will now be developed in a general thermodynamic framework in Section 3. Before this is done, however, the thermodynamic foundations are laid in Section 2.

SECTION 2

Thermodynamic Foundations.

The following are the fundamental laws of thermodynamics, which apply to all continuous media irrespective of their constitutive properties. (For materials that are solid-like, in the sense that they have a memory of their initial configuration, it is more convenient to express these laws in the material coordinate system x^i). In differential form, these are the first law of thermodynamics,

$$\dot{\epsilon} = (\rho_0/2\rho)\, \tau^{ij}\, \dot{C}_{ij} - h^i{}_{,i} + \dot{Q}, \tag{2.1}$$

the rate of dissipation inequality,

$$\theta\dot{\gamma} = (\rho_0/2\rho)\, \tau^{ij}\, \dot{C}_{ij} - \dot{\psi} - \eta\dot{\theta} \geq 0 \tag{2.2}$$

and the heat conduction inequality

$$-h^i\theta_{,i} \geq 0. \tag{2.3}$$

The symbols in eq.'s (2.1), (2.2) and (2.3) have the following meaning : ϵ is the internal energy per unit mass : ρ_0 and ρ are the initial and current mass densities respectively : τ^{ij} is the stress tensor in the material coordinate system x^i ; C_{ij} is the right Cauchy-Green tensor, h^i is the heat flux vector per unit undeformed area in the material system ; Q is the heat supply per unit mass; θ is the temperature, γ the irreversible entropy and ψ and η are the free energy and entropy, respectively, per unit underformed volume finally a subscript following a comma denotes differentiation with respect to the corresponding material coordinate. A dot over a quantity denotes material derivative with respect to time. To avoid repetitious statements, henceforth we shall refer to C_{ij} as the "deformation".

In the case of dissipative materials the stress, the internal energy and entropy densities (and, therefore, the free energy density) of a material neighborhood depend on the entire history of deformation and temperature of that neighborhood.

In the theory of irreversible thermodynamics the effects of history are taken into account by specifying that the stress and free energy density are functions of the current values of C_{ij} and θ as well as n additional independent variables q_a, not necessarily observable, called "internal variables". These may be scalars or components of vectors or tensors in the material frame ; whatever their geometric nature they must remain invariant with translation and rotation of the spatial system to satisfy the principle of material indifference. Thus :

$$\psi = \psi\,(C_{ij}, \theta, q_a) \tag{2.4}$$

$$\tau^{ij} = \tau^{ij}\,(C_{k\ell}, \theta, q_a) \tag{2.5}$$

It has been shown elsewhere (4) that

$$\tau^{ij} = \frac{2\rho}{\rho_o} \frac{\partial \psi}{\partial C_{ij}} \tag{2.6}$$

$$\eta = -\frac{\partial \psi}{\partial \theta} \tag{2.7}$$

$$\theta \dot{\gamma} = -\frac{\partial \psi}{\partial q_a} \dot{q}_a \geq 0 \tag{2.8}$$

Furthermore the heat flux vector h^i is a function of $\theta_{,i}$, θ, C_{ij} and q_a i.e.,

$$h^i = h^i\,(\theta_{,i}, \theta, C_{k\ell}, q_a) \tag{2.9}$$

subject to the conditions :

(2.10a,b) $$h^i\big|_{\theta,i=0} = 0, \quad h^i\theta_{,i} \leq 0.$$

Finally eq. (2.1) in conjunction with eq.' (2.6) and (2.7) yields

(2.11) $$\dot{h}^i_{,i} = \theta\left(\frac{\partial\psi}{\partial\theta}\right)^{\cdot} - \frac{\partial\psi}{\partial q_a}\dot{q}_a + \dot{Q}$$

The remarkable property of the above equations is that they apply to all materials irrespective of their constitution. This has not been generally recognized. In fact the constitutive nature of the material follows from the constitutive properties of q_a. For example, in elastic materials $q_a = 0$, whereas in viscoelastic materials q_a are given by a set of differential equations of the type,

(2.12) $$\dot{q}_a = f_a\,(q_\beta\,,\,c_{ij}\,,\,\theta)$$

The question of how q_a are determined for viscoplastic materials is considered in the next Section.

SECTION 3

Constitutive Equations in Viscoplasticity.

From the right hand side of eq. (2.8) and the fact that $\frac{d\zeta}{dt} > 0$ and $dz/d\zeta > 0$, it follows that

$$- \frac{\partial \psi}{\partial q_a} \frac{dq_a}{dz} \geq 0 \text{ (} a \text{ not summed)} \tag{3.1}$$

where the inequality is valid unless $\frac{dq_a}{dz} = 0$. It also follows from inequality (3.1) that $\frac{dq_a}{dz}$, q_a, C_{ij} and θ must be related otherwise $\frac{\partial \psi}{\partial q_a}$ and $\frac{dq_a}{dz}$ could be prescribed independently and in such a fashion, that the inequality would be violated. In this event there must exist a set of relations

$$\frac{dq_a}{dz} = f_a (C_{ij}, q_\beta, \theta) \tag{3.2}$$

for all a, where the f_a are material functions.

It must be noted that, as a result of eq. (3.2) q_a are indeed functionals of the histories of deformation and temperature with respect, however, to the intrinsic time scale z which is, itself, a material property.

Thus, at least formally, the constitutive equations of the endochronic theory of viscoplasticity are now complete in the sense that given the material function f_a, ψ and h^i then for some specified deformation and temperature histories, q_a are found from eq. (3.2) and thus τ^{ij} and η are found from eq.'s (2.6) and (2.7) respectively ; similarly h^i the heat flux vector is determined from eq. (2.9).

Ideally, one would like to know the thermomechanical three-dimensional response of a material over the whole spectrum of mechanical and thermal conditions, i.e., under all variations in strain, strain rate (or more generally, history of strain) and temperature. However, such a task would be a momentous, if not an impossible, undertaking ; the experimental evaluation of the material functions involved, under wide conditions of strain and temperature, would be impractical.

Fortunately the domain of specification of design conditions is usually limited in some way ; for instance usually, (a) large changes of temperature, fast rates of loading, but small strains are prescribed ; or (b) small changes in temperature and small rates of loading but large strains and/or displacements prevail. More extreme mechanical as well as thermal conditions are rarer.

It is reasonable to expect that material behavior would be easier to describe mathematically over a narrower domain of environmental conditions, where the applicability or "correctness" of such mathematical formulation would be easier to check experimentally.

In what follows we shall consider situations in which the strain in a material region R as well as the temperature changes relative to a uniform reference temperature θ_0 are "small". To make the above statement more precise let $\epsilon_{ij}(z')$ denote the history of the strain tensor (*), for $z_0 < z' < z$ where z_0 is some initial intrinsic time. Set

$$\|\epsilon_{ij}(z')\| \equiv \{\epsilon_{ij}(z')\,\epsilon_{ij}(z')\}^{1/2} \tag{3.3}$$

(*) $\epsilon_{ij} = {}^1/_2\,(C_{ij} - \delta_{ij})$

and let the supremum of $||\epsilon_{ij}(z')||$ be Δ.

Similarly let $\vartheta(z')$ be the history of the temperature increment relative to the reference temperature θ_0 and let the supremum of $|\vartheta(z')|$ be δ. The notion of smallness is made precise by stipulating that $\Delta << 1$ and $\delta << 1$.

Thus, formally

$$\theta = \theta_o + \vartheta \tag{3.4}$$

$$\eta = \eta_o + \chi \tag{3.5}$$

$$||\epsilon_{ij}(z')||_{sup} = \Delta, \ |\vartheta(z')|_{sup} = \delta \tag{3.6}$$

To complete the formalism let χ be the entropy change relative and a reference uniform entropy η_0, and let σ_{ij} denote the stress tensor. The reference state is defined by the condition that $\sigma_{ij} = 0$, $\psi = \vartheta = \chi = 0$, $q_a = 0$.

Under these conditions eq.'s (2.4) through (2.9) and eq. (2.10b) become,

$$\psi = \psi\,(\epsilon_{ij}, \vartheta, q_a) \tag{3.7}$$

$$\sigma_{ij} = \frac{\partial\psi}{\partial\epsilon_{ij}} \tag{3.8}$$

$$\chi = -\frac{\partial\psi}{\partial\vartheta} \tag{3.9}$$

$$\theta_0\dot{\gamma} = -\frac{\partial\psi}{\partial q_a}\,\dot{q}_a \geq 0 \tag{3.10}$$

$$h_i = k_{ij}\vartheta_{,j} \tag{3.11}$$

$$h_i\vartheta_{,i} \leq 0 \tag{3.12}$$

where k_{ij} is the thermal conductivity tensor.

Finally eq. (2.11), becomes

$$h_{i,i} = \theta_o \left(\frac{\partial \psi}{\partial \vartheta} \right)^{\cdot} - \frac{\partial \psi}{\partial q_a} \dot{q}_a + \dot{Q}. \tag{3.13}$$

It is shown in Appendix II that $|q_a|$ and $|q_a|$ may stay small in the sense that given two positive numbers δ_1^a and δ_2^a , however small, Δ and δ can be chosen small enough such that $|q_a| < \delta_1^a$ and $|q_a| < \delta_2^a$.

At this stage one may obtain the corresponding equation for q_a by linearizing eq. (3.2). However it is physically more meaningful and, a posteriori, more rewarding to examine more closely the rate of change of irreversible entropy γ.

From eq. (3.10),

$$\theta_o \frac{d\gamma}{dz} = - \frac{\partial \psi}{\partial q_a} \frac{dq_a}{dz} \geq 0 \tag{3.14}$$

It follows from eq.'(s) (3.14) and (3.2) that $\frac{d\gamma}{dz}$ may be expressed as a function of $\frac{dq_a}{dz}$, ϵ_{ij} and ϑ, subject to the condition that $\frac{d\gamma}{dz} = 0$ whenever $\frac{dq_a}{dz} = 0$ for all a. Thus, if we expand $\theta_o \frac{d\gamma}{dz}$ in a Taylor series and ignore terms of order higher than $0(\hat{\delta}^2)$(*) and observe the inequality (3.10), thereby eliminating the linear terms in the expansion,then

$$\theta_o \frac{d\gamma}{dz} = b_{a\beta} \frac{dq_a}{dz} \frac{dq_\beta}{dz} \tag{3.15}$$

Eq.'s (3.14) and (3.15) are simultaneously satisfied if

$$\frac{\partial \psi}{\partial q_a} + b_{a\beta} \frac{dq_a}{dz} = 0 \tag{3.16}$$

(*) δ_a is the largest of δ_1^a and δ_2^a. Also, $\hat{\delta}$ is the largest of δ, Δ and δ_1^a.

With eq. (3.16) the constitutive description of a viscoplastic material is now complete.

Explicit Constitutive Equations.

Explicit constitutive equations for viscoplastic materials under conditions of small strain and small changes in temperature are obtained by expanding ψ in eq. (3.7) in Taylor Series and omitting terms of order higher than $0(\hat{\delta}^2)$; (*) linear terms must vanish to satisfy the initial conditions.

Before the expansion is carried out, however, it appears desirable to regard q_a not as scalars but components of second order tensors. This, as will be shown, obviates certain difficulties which arise with the representation of fourth order tensors. For instance, in Ref. (4), we were faced with having to assume, without proof, that a fourth order tensor $C_{ijk\ell}$ such that,

$$C_{ijk\ell} = C_{jik\ell} = C_{ij\ell k} = C_{k\ell ij}$$

is given by the series

$$C_{ijk\ell} = \sum_a \frac{a_{ija} a_{k\ell a}}{a_a} \tag{3.17}$$

where a_{ija} are second order symmetric tensors and a_a are scalars. Problems such as this are obviated by giving the internal variables a tensorial character. Thus the free energy density and other thermodynamic quantities are now functions of ϵ_{ij}, ϑ and n internal variables q_{ij}^a $(a = 1, 2 \ldots . n)$, where q_{ij}^a are symmetric second order tensors, with respect to the material system x_i. In this notation, eq.'s (3.13)

(*) δ_a is the largest of δ_1^a and δ_2^a. Also, $\hat{\delta}$ is the largest of δ, Δ and δ_1^a.

through (3.16) now read,

(3.18) $$h_{i,i} = \theta_0 \frac{\partial \psi^{\cdot}}{\partial \vartheta} - \frac{\partial \psi}{\partial q^{a}_{ij}} \dot{q}^{a}_{ij} + \dot{Q}$$

(3.19) $$\theta_0 \frac{d\gamma}{dz} = - \frac{\partial \psi}{\partial q^{a}_{ij}} \frac{dq^{a}_{ij}}{dz} \geq 0$$

(3.20)(*) $$\theta_0 \frac{d\gamma}{dz} = \sum_{a} b^{a}_{ijk\ell} \frac{dq^{a}}{dz}{}_{ij} \frac{dq^{a}}{dz}{}_{k\ell}$$

and

(3.21) $$\frac{\partial \psi}{\partial q^{a}_{ij}} + b^{a}_{ijk\ell} \frac{dq^{a}_{k\ell}}{dz} = 0$$

(a not summed)

Furthermore, in view of my previous discussion,

(3.22) $$\psi = \tfrac{1}{2} A_{ijk\ell}\, \epsilon_{ij}\, \epsilon_{k\ell} + B^{a}_{ijk\ell}\, \epsilon_{ij}\, q^{a}_{k\ell} + \sum_{a} C^{a}_{ijk\ell}\, q^{a}_{ij}\, q^{a}_{k\ell} + D_{ij} \vartheta \epsilon_{ij} + E^{a}_{ij} \vartheta q^{a}_{ij} + \tfrac{1}{2} F \vartheta^2 .$$

Though, in principle, eq.'s (3.8), (3.9), (3.18), (3.21) and (3.22) are sufficient for the derivation of explicit constitutive equations, we shall obtain these only for isotropic material, so as to keep the algebra at a minimum. For such materials

(3.23a) $$A_{ijk\ell} = A_1\, \delta_{ij}\, \delta_{k\ell} + A_2\, \delta_{ik}\, \delta_{j\ell}$$

(*) Expansions of the type $b^{a\beta}_{ijk\ell} \frac{dq^{a}_{ij}}{dz} \frac{dq^{\beta}_{k\ell}}{dz}$ and $A^{a\beta}_{ijk\ell}\, q^{a}_{ij}\, q^{\beta}_{k\ell}$ reduce to the above form. See Ref. 5.

$$B^{a}_{ijk\ell} = B^{a}_{1}\ \delta_{ij}\ \delta_{k\ell} + B^{a}_{2}\ \delta_{ik}\ \delta_{j\ell}$$

$$C^{a}_{ijk\ell} = C^{a}_{1}\ \delta_{ij}\ \delta_{k\ell} + C^{a}_{2}\ \delta_{ik}\ \delta_{j\ell}$$

$$D_{ij} = D\ \delta_{ij}\ ,\ E^{a}_{ij} = E^{a}\ \delta_{ij}$$

$$k_{ij} = k\ \delta_{ij}\ ,\ b^{a}_{ijk\ell} = b^{a}_{1}\ \delta_{ij}\ \delta_{k\ell} + b^{a}_{2}\ \delta_{ik}\ \delta_{j\ell} \tag{3.23b-f}$$

It is worth noting that here we consider materials which are "stable" in the sense that straining of the reference configuration under isothermal conditions will increase the free energy density ψ. Thus $A_{ijk\ell}$ and $C^{a}_{ijk\ell}$ are positive definite. As a consequence A_0, A_2, C^{a}_{0} and C^{a}_{2} are all positive (*). Furthermore, as a result of eq. (3.20) and inequality (3.19) $b^{a}_{ijk\ell}$ is a positive definite tensor, in which case b^{a}_{0} and b^{a}_{2} are positive, in the notation of eq. (3.33).

Omitting superfluous algebra, the coupled thermomechanical constitutive equations take the following form in terms of the hydrostatic stress $\sigma = \frac{\sigma_{kk}}{3}$, the deviatoric stress tensor σ_{ij}, the increment in temperature, the hydrostatic strain ϵ_{kk}, the deviatoric strain tensor e_{ij} and the entropy increment χ ; in terms of the above notation :

$$s_{ij} = 2 \int_{z_o}^{z} \mu(z-z') \frac{\partial e_{ij}}{\partial z'} dz' \tag{3.24}$$

$$\sigma = \int_{z_o}^{z} K(z-z') \frac{\partial z_{kk}}{\partial z'} dz' + \int_{z_o}^{z} D(z-z') \frac{\partial \vartheta}{\partial z} dz \tag{3.25}$$

(*) In the notation of eq. (3.31).

(3.26) $$-\chi = \frac{\partial\psi}{\partial\vartheta} = \int_{z_0}^{z} D(z-z')\frac{\partial\epsilon_{kk}}{\partial z'}\,dz' + \int_{z_0}^{z} F(z-z')\frac{\partial\vartheta}{\partial z'}\,dz'$$

where

(3.27) $$2\,\mu(z) = \left(A_2 - \sum_a \frac{B_2^a\,B_2^a}{C_2^a}\right) H(z) + \sum_a \frac{B_2^a\,B_2^a}{C_2^a}\, e^{-\rho_a z}$$

(3.28) $$K\,(z) = \left(A_0 - \Sigma \frac{B_0^a\,B_0^a}{C_0^a}\right) H(z) + \sum_a \frac{B_0^a\,B_0^a}{C_0^a}\, e^{-\lambda_a z}$$

(3.29) $$D\,(z) = \left(D - \sum_a \frac{B_a\,E_a}{C_0^a}\right) H(z) + \sum_a \frac{B_a\,E_a}{C_0^a}\, e^{-\lambda_a z}$$

(3.30) $$F\,(z) = \left(F - \sum_a \frac{E^a\,E^a}{C_0^a}\right) H(z) + \sum_a \frac{E^a\,E^a}{C_0^a}\, e^{-\lambda_a z}$$

$$A_0 = 1/3\,(3A_1 + A_2),\quad B_0 = 1/3\,(3B_1 + B_2),\quad C_0^a = 1/3\,(3C_1^a + C_2^a)$$

(3.31)

(3.32) $$\rho_a = \frac{C_2^a}{b_2^a}\,,\quad \lambda_a = \frac{C_0^a}{b_0^a}$$

(3.33) $$b_0^a = 3b_1^a + b_2^a$$

The heat conduction equation is similarly found to be :

$$\frac{dt}{dz} k\, \vartheta_{,ii} = \frac{\partial}{\partial z} \int_{z_o}^{z} C_v(z-z') \frac{\partial \vartheta}{\partial z'} dz' - \theta_o \frac{\partial}{\partial z} \int_{z_o}^{z} D(z-z') \frac{\partial z_{kk}}{\partial z'} dz'$$

$$+ \hat{Q} + \sum_a b_1^a \hat{q}_{ii}^a \hat{q}_{jj}^a + \sum_a b_2^a \hat{q}_{ij} \hat{q}_{ij} \qquad (3.34)$$

where

$$C_v(z) = -\theta_o F(z) \qquad (3.35)$$

and a roof over a quantity implies differentiation with respect to z. The lower limit z_0 denotes the intrinsic time of the reference state.

SECTION 4

Endochronic Theory of Plasticity and its Relation to Present Theories

Our theory of plasticity, which is a rate independent endochronic theory, is obtained by replacing the time measure $d\zeta$ by $d\xi$. The time scale now becomes $z(\xi)$, but the form of the constitutive equations remains unaltered. In particular, the "linear" form of our theory is obtained by setting

$$d\xi^2 = P_{ijk\ell}\, d\epsilon_{ij}\, d\epsilon_{k\ell} \tag{4.1}$$

where $P_{ijk\ell}$ is a positive definite fourth order material tensor. We repeat the constitutive equations of the linear theory, in the particular case when the deformation is isothermal so that a comparison may be made with current theories. When $\vartheta \equiv 0$, then eq.'s (3.24) and (3.25) become,

$$s_{ij} = 2 \int_{z_o}^{z} \mu(z-z') \frac{\partial e_{ij}}{\partial z'}\, dz' \tag{4.2}$$

$$\sigma_{kk} = 3 \int_{z_o}^{z} K(z-z') \frac{\partial \epsilon_{kk}}{\partial z'}\, dz' \tag{4.3}$$

where $z = z(\xi)$.

If the material behaves elastically under pressure (so-called plastically incompressible) then $K(z)$ is a constant and in this case

$$\sigma_{kk} = 3K\epsilon_{kk} \tag{4.4}$$

Whereas,

$$s_{ij} = 2 \int_{z_o}^{z} \mu(z-z')\, de_{ij}(z') \tag{4.5}$$

Let now $\mu(z)$ consist of a single exponential term i.e.

$$\mu(z) = \mu_0 \, e^{-az} \tag{4.6}$$

In this event

$$s_{ij} = 2\mu_0 \int_{z_0}^{z} e^{-a(z-z')} \, de_{ij}(z') \tag{4.7}$$

The integral eq. (4.7) is reducible to the differential equation

$$de_{ij} = \frac{a}{2\mu_0} dz \; s_{ij} + \frac{1}{2\mu_0} ds_{ij} \tag{4.8}$$

Now, $\frac{1}{2\mu_0} ds_{ij}$ may be identified as the "elastic" component of deviatoric strain of classical plasticity. If one follows the traditional definition of "plastic strain" $de_{ij}P$ given below, i.e.,

$$de_{ij}P = d\epsilon_{ij} - d\epsilon_{ij}^{e} \tag{4.9}$$

then in view of eq. (48)

$$de_{ij}P = \frac{a}{2\mu_0} dz \; s_{ij} \tag{4.10}$$

But these are the Prandtl-Reuss relations. Hence our present theory contains these relations as a special case. Where then does it differ from this theory ? It does in the interpretation of the proportionality coefficient dz. In the Prandtl-Reuss theory $d\xi$ may be positive negatice or zero, and in fact, z has been identified with the yield surface, i.e., plastic action is assumed to occur when $dz < 0$, where

$$z = z(s_{ij}) \tag{4.11}$$

but that de_{ij}^P is zero whenever

$$dz \leq 0 \tag{4.12}$$

In the present theory dz is always positive if the material is deforming (it is zero only when deformation does not take place). Thus always,

$$dz \geq 0 \tag{4.13}$$

Furthermore dz is not given by eq. (4.11) i.e. it is not related to some yield surface but its definition is entirely kinematic. Thus, no yield phenomenon or surface are postulated here. One obtains the stress response by merely monitoring the history of strain.

Also the theory admits a further generality since $\mu(z)$ need not consist of a single exponential term.

For instance $\mu(z)$ may be of the form

$$\mu(z) = \mu_0 + \mu_1 e^{-az} . \tag{4.14}$$

In this case, however, the differential form of eq. (4.7) becomes :

$$2(\mu_0 + \mu_1)\, de_{ij} + 2\mu_0\, a\, e_{ij}\, dz = ds_{ij} + a\, s_{ij}\, dz. \tag{4.15}$$

The shear modulus μ_I, at $z = 0$, (initial modulus), is $(\mu_0 + \mu_1)$. The "plastic" components of the deviatoric shear strain tensor are given from eq. (4.9), i.e..,

$$de_{ij}^P = \frac{a}{2\mu_I}\, dz\, s_{ij} - 2\,\mu_0\, e_{ij}. \tag{4.16}$$

Note that eq. (4.16) does not satisfy the Prandtl-Reuss relations, which are also violated if one adds more exponential terms to the right hand side of eq. (4.14). In fact these relations will be satisfied if and only if $\mu(z)$ is given by eq. (4.6), i.e., μ is represented by a single exponential term only. This situation is not particularly disturbing. Peters et Als (40) carried out experiments on thin walled 14S-T4 aluminum alloy cylinders by loading these in combined compression and torsion and found that the Prandtl-Reuss relations were not satisfied, for this particular metal.

CONCLUSIONS

A theory has been presented here, the scope of which is wide enough to allow a rational phenomenological description of mechanical behavior of materials under various histories of strain and temperature. In particular, the viscoplastic behavior of materials is formulated mathematically, without recourse to the dichotomy of the deformation history in plastic and elastic parts and without the necessity of introducing discontinuities in material behavior, such as yield surfaces.

The theory merely asserts that, to every history of deformation gradient and temperature of a neighborhood there corresponds a unique state of stress in that neighborhood. An entirely novel feature of the theory is that these histories are defined with respect to a time scale, which itself is a material property.

In this paper, we have merely presented the framework of the theory without actually evaluating the material functions involved, through the use of experimental data. This, however, will be done in Part II of this paper, where it will be shown that the theory describes experimentally observed plastic behavior of metals with a remarkable degree of accuracy.

Appendix I

The following is a short account of the experimental work on (a) the effect of the definition of yield on the shape of the yield surface and (b) of the work on viscoplasticity. The references given are by no means exhaustive and the author wishes to apologize to people of whose work he is not currently aware.

In Ref.'s 6 and 7, aluminum alloy tubes (**) were subjected to shear prestrain by twisting well into the plastic region by a predetermined amount. The yield surface corresponding to this degree of prestrain was established by loading the tubes in combined tension and torsion.

In Ref. 6, Naghdi found that subsequent yield surfaces distorted in the direction of the shear axis with a pronounced Bauschinger effect in shear but there was no effect on the yield stress in tension (i.e. the yield locus did not change in the vicinity of zero shear stress).

On the other hand, in Ref. 7, Ivey observed that the yield surface, in addition to distortion, underwent a large amount of translation in the direction of of the shear axis, so that for large prestrains the origin of the stress space was outside the yield surface. However, he was in agreement with Naghdi in that the presence of shear prestrain did not aggect the yield stress in tension. Both authors used deviation from kinearity in a stress strain diagram as a definition of yield.

Mair and Pugh (8) to check the absence of "cross effect", carried out their own experiments on copper with a high degree of isotropy. However they used a different definition of yield, this being the point of intersection of the initial straight part of the stress-strain curve with a backward linear extrapolation of the "plastic" part of the stress-strain curve.

Their results varied significantly from those of Ivey and Naghdi. They found that expansion and istortion of the initial locus took place with a strong cross effect between shear and tension. Also a pronounced Bauschinger effect in torsion was found with large initial positive pretorsion. These authors also observed pronounced "plastic" unloading in shear.

The results of Szczepinski and Miastkowski (9) tend to confirm the findings of Mair and Pugh (8). Their results, moreover, were significant in other respects. Specifically, using the proof strain to define yield, they studied aluminum alloy sheets under biaxial tension with the intention of finding the effect of prestrain on the shape of the yield surfaces. They observed, migration, distortion, expansion and sometimes rotation of the initial yield locus.

Similar conclusions (*) can be drawn from Szczepinski's paper (10), as well as Miastkowski and Szczepinski's (17), in which tubular brass specimens were subjected to combined axial and circumferential stress.

Initial and subsequent yield loci were plotted when yield was defined (a) as departure from kinearity or (b) when it was set to correspond to a certain proof strain. In particular, when definition (a) was used, subsequent yield loci did not contain the initial locus, but when (b) was used, with proof strain set at 0.5 subsequent loci contained the initial locus.

Attempts to describe the change of the yield locus with prestrain, by simple models have not proved satisfactory. Batdorf and Budianski (13), suggested that after prestrain, the yield locus is the minimum surface through the point of

---

(*) In this connection, see also work of the same general nature by Bertsch and Findley (11) and Hu and Bratt (12).

prestrain and the initial yield locus. This model however does not account for the Bauschinger effect. The kinematic hardening rulle (*) proposed by Prager (14), was partially successful, in so far as it can be of value only when the stress-strain curve of a material in simple tension is bilinear (15). Otherwise subsequent shapes of yield locus must be defined in terms of a parameter that depends on the history of strain (15) to obtain realistic unloading behavior.

A more realistic model is the one by Hodge (16) which included translation, expansion and distorsion of the yield surface. This model covers all contingencies but does not include the history of stress on the shape and position of the yield locus.

However, every definition of yield gives rise to a different yield surface. If we insist that the increment of plastic strain is to be normal to the yield surface, then, for a complex but specific loading history, each such definition will give rise to a different plastic strain history. Only one of these can be the correct one.

So it appears that through Eisenberg's and Phillip's (15) mathematical description of a yield surface has been most promising, we must be prepared to question, if necessary, whether the concept of yield point and yield surface are the only way by which plastic effects may be described, especially in view of the fact that these may take place immediately following the initiation of deformation of material, though they may be negligible in the region of small strains. This would agree with the point of view that dislocations (and, therefore, plastic behavior) originate immediately upon initiation of the loading.

(see also Ref. (19).

Viscoplasticity.

The need for the development of the theory of viscoplasticity arises from the recognition of the strain rate sensitivity of metals under dynamic loading.

The difficulty in trying to synthesize a rational "rate" theory from experimental observations, a priori, lies in the fact that under dynamic conditions the inertia effects are significant. In the absence of a constitutive theory, these effects cannot be calculated (*). Therefore, in the case of dynamic theories, such as viscoplasticity, theory and experiments must advance together.

The literature abounds with data on the subject of strain rate sensitivity, particularly in one dimension (20-30). Lindholm (32) carried out dynamic experiments in one and two dimensions in an attempt to generalize results which were arrived at, by consideration of thermally activated processes and their relation to dislocation theory in metals. See Ref.'s 35-40.

An early attempt at a theoretical viscoplastic constitutive equation in one dimension is due to Malvern (31, 32). This equation assumes the existence of a "static" stress-strain relation and then relates the stress increment, with respect the static value, to the strain rate.

Modifications and generalizations of Malvern's equation were made by Lubliner (33) who included a limiting maximum stress-strain curve, and by Perzyna (34), Perzyna and Wojno (35) who proposed a multiaxial generalization for finite strains, assuming the additivity of the elastic and plastic strain components

(*) Constant strain rate experiments would appear to be an exception, by being less susceptible to inertia effects. However, Ref. 38 tends to negate this. Long specimens give different responses to short ones, under the same conditions.

and by Perzyna (39) who used concepts of internal coordinates and irreversible thermodynamics to eliminate the above assumptions and to put the theory on sounder foundations.

Though, in his last treatment, Perzyna (39) abandoned the additivity of plastic and elastic strains, he still retained the concepts of yield stress (and yield surface) and the hypothesis of a datum plastic stress strain relation, with respect to which "strain history" is to be related to the "excess" stress through the internal coordinates.

Our theory differs from Perzyna's theory in this respect.

We close by mentioning that, with the exception of the papers by Perzyna (39), only a moderate research effort has been made in the area of coupling between a viscoplastic and a thermal process. However, Chidister and Malvern (25), Lindholm (27) and Trozera, Sherby and Dorn (37), considered the effect of a change in uniform temperature on viscoplastic behavior, with a view to confirming some results of the dislocation theory.

Appendix II

The form of eq. (3.22) for isotropic materials, according to eq.'s (3.22 a-e), is

$$\psi = \tfrac{1}{2}\, A_1\, \epsilon_{ii}\, \epsilon_{jj} + \tfrac{1}{2}\, A_2\, \epsilon_{ij}\, \epsilon_{ij} + \overset{a}{B}_1 \epsilon_{ii} \overset{a}{q}_{jj} + \overset{a}{B}_2\, \epsilon_{ij}\, \overset{a}{q}_{ij}$$

$$+ \tfrac{1}{2}\, \overset{a}{C}_1\, \overset{a}{q}_{ii}\, \overset{a}{q}_{jj} + \tfrac{1}{2}\, \overset{a}{C}_2\, \overset{a}{q}_{ij}\, \overset{a}{q}_{ij} + D\vartheta\epsilon_{jj} + \overset{a}{E}\, \vartheta \overset{a}{q}_{ii} + \tfrac{1}{2}\, F\vartheta^2 \quad \text{(A.2.1)}$$

As a result eq.'s (3.8) and (3.9) yield :

$$\sigma_{ij} = A_1\, \delta_{ij}\, \epsilon_{kk} + A_2 \epsilon_{ij} + \overset{a}{B}_1\, \epsilon_{ij}\, \overset{a}{q}_{kk} + \overset{a}{B}_2\, \overset{a}{q}_{ij} + D\vartheta\delta_{ij} \quad \text{(A.2.2)}$$

$$(a \text{ summed})$$

$$-\chi = \frac{\partial \psi}{\partial \vartheta} = D\epsilon_{ii} + \overset{a}{E}\, \overset{a}{q}_{ii} + F\vartheta \ (a \text{ summed}) \quad \text{(A.2.3)}$$

On the other hand,

$$\frac{\partial \psi}{\partial \overset{a}{q}_{ij}} = \overset{a}{B}_1\, \epsilon_{kk}\, \delta_{ij} + \overset{a}{B}_2\, \epsilon_{ij} + \overset{a}{C}_1\, \delta_{ij}\, \overset{a}{q}_{kk}$$

$$+ \overset{a}{C}_2\, \overset{a}{q}_{ij} + \overset{a}{E}\, \delta_{ij}\vartheta \ (k \text{ not summed}) \quad \text{(A.2.3)}$$

Hence, use of eq. (3.21) in accordance with eq. (3.23f) yields a set of first order differential equations for $\overset{a}{q}_{ij}$; these can be expressed as a set for $\overset{a}{q}_{kk}$ and another for the deviatoric part of $\overset{a}{q}_{ij}$, which we denote by $\overset{a}{p}_{ij}$. Thus in the notation of eq.'s (3.31) and (3.33)

$$\text{(A.2.4)} \qquad \overset{a}{B_0}\,\epsilon_{kk} + \overset{a}{C_0}\,\overset{a}{q_{kk}} + \overset{a}{E} + \overset{a}{b_0}\,\frac{d\overset{a}{q_{kk}}}{dz} = 0$$

$$\text{(A.2.5)} \qquad \overset{a}{B_2}\,e_{ij} + \overset{a}{C_2}\,\overset{a}{p_{ij}} + \overset{a}{b_2}\,\frac{dp_{ij}}{dz} = 0$$

In both eq.'s (A.2.4) and (A.2.5) a is not summed. It follows from the above two equations that

$$\text{(A.2.6)} \qquad \overset{a}{q_{kk}} = -\frac{\overset{a}{B_0}}{b_0^a}\int_{z_0}^{z} e^{-\lambda_a(z-z')}\,\epsilon_{kk}(z')dz' - \frac{\overset{a}{E}}{b_0^a}\int_{z_0}^{z} e^{-\lambda_a(z-z')}\,\vartheta(z')dz'$$

$$\text{(A.2.7)} \qquad \overset{a}{p_{ij}} = -\frac{\overset{a}{B_2}}{b_2^a}\int_{z_0}^{z} e^{-\rho_a(z-z')}\,e_{ij}(z')dz'$$

where λ_a and ρ_a are given by eq. (3.32).

In the light of the tensorial notation that we have adopted for the internal variables, let

$$\text{(A.2.8)} \qquad \left\|\epsilon_{ij}\right\|_{sup} = \Delta, \quad \left\|e_{ij}\right\|_{sup} = \Delta_1\,, \quad \left|\epsilon_{kk}\right|_{sup} = \Delta_0$$

where, $\|\epsilon_{ij}\| = |\epsilon_{ij}\,\epsilon_{ij}|^{1/2}$, etc. Evidently,

$$\text{(A.2.9)} \qquad \Delta^2 = \Delta_1^2 + \frac{1}{3}\Delta_0^2$$

Then as a result of eq.s (A.2.7) and (A.2.8)

(A.2.10) $$\left\| \overset{a}{p}_{ij} \right\| \leq \frac{|\overset{a}{B}_2|}{\overset{a}{C}_2} \Delta_1$$

(A.2.11) $$|\overset{a}{q}_{kk}| \leq \frac{|\overset{a}{B}_0|}{\overset{a}{C}_0} \Delta_0 + \frac{|\overset{a}{E}|}{\overset{a}{C}_0} \delta$$

where as before $|\vartheta|_{sup} = \delta$

Also from eq. (A.2.5),

$$(\overset{a}{b}_2)^2 \left\| \frac{dp_{ij}}{dz} \right\|^2 = (\underset{a}{B}^2)^2 \left\| e_{ij} \right\|^2 + (\underset{a}{C}^2)^2 \left\| p_{ij} \right\|^2$$

(A.2.12) $$+ 2B_a^2 \, C_a^2 \left| p_{ij} \, e_{ij} \right|$$

However, since

(A.2.13) $$\left| p_{ij} \, e_{ij} \right| \leq \left\| p_{ij} \right\| \left\| e_{ij} \right\|$$

it follows from (A.2.12) that

(A.2.14) $$\overset{a}{b}_2 \left\| \frac{dp_{ij}}{dz} \right\| \leq 2 \, \Delta_1 \left| \overset{a}{B}_2 \right|$$

Also as a result of eq. (A.2.4)

(A.2.15) $$\overset{a}{b}_0 \left| \frac{d\overset{a}{q}_{kk}}{dz} \right| \leq 2 \left| \overset{a}{B}_2 \right| \Delta_0 + 2 \left| E^a \right| \delta$$

At this point we order our internal variables as shown,

$$\overset{1}{p}_{ij} , \overset{2}{p}_{ij} \cdot \cdot \cdot \cdot \cdot \overset{m}{p}_{ij} ; \overset{1}{q}_{kk} , \overset{2}{q}_{kk} \cdot \cdot \cdot \cdot \overset{m}{q}_{kk} \cdot$$

Let q_a be a typical internal variable. Then, whether it belongs to the p-group or the q-group above, as a result of eq.'s (A.2.10), (A.2.11), (A.2.14) and (A.2.14), given two positive members δ_1 and δ_2, however small, we can choose Δ_0 and Δ_1 (and therefore Δ) and δ such that

$$|q_a| \leq \delta_1 \text{ and } \left|\frac{dq_a}{dz}\right| \leq \delta_2 .$$

REFERENCES

1. Ilyushin A. A., "On the relation between stresses and small deformations in the mechanics of continuous media," Prikl. Math. Mech., 18, 641, (1954).

2. Rivlin R. S., "Nonlinear viscoelastic solids," SIAM Review 7, 323, (1965).

3. Schapery, R. A., "On a thermodynamic constitutive theory and its application to various non-linear materials," Proceedings, IUTAM Symposium, East Kilbride, 259, (1968).

4. Valanis, K. C., "A unified theory of thermomechanical behavior of viscoelastic materials," Mechanical Behavior of Materials Under Dynamic Loads, Ed. U.S. Lindholm, Springer-Verlag, N. Y. (1968).

5. Valanis, K. C. "Thermodynamics of large viscrelastic deformations, J. Math and Phys., 45, 197, (1966).

6. Naghdi, P. M., Essenburgh, F. and Koff, W., "An experimental study of initial and subsequent yield surfaces in plasticity," J. App. Mech., 25, 201, (1958).

7. Ivey, H. J., "Plastic stress-strain relations and yield surfaces for aluminum alloys," J. Mech. Eng. Sc., 3, 15, (1961).

8. Mair W. M. and Pugh, H. L. 1. D., "Effect of prestrain on yield surfaces in copper," J. Mech. Eng. Sc., 6, 150 (1964).

9. Szczepinski, W. and Miastkowski, J., "An experimental study of the prestraining history on the yield surfaces of an aluminum alloy," J. Mech. ph. Sol. 16, 153, (1968).

10. Szczepinski, W. "On the effect of plastic deformation on the yield condition," Bulletin de l'académie polonaise des sciences, Série des sciences techniques, 11, 463, (1963).

11. Bertsch, P. K. and Findley, W. N., "An experimental study of subsequent yield surfaces, normality, Bauschinger and allied effects," Proc. 4th U. S. Congr. App Mech., Berkley (1962).

12. Hu, L. W. and Bratt, J. F., J. App. Mech. 25, 441 (1958).

13. Batdorf, S. F. and Budianski, B., "Polyaxial stress-strain relations of strain-hardening metals," J. App. Mech., 21, 323, (1954).

14. Prager, W., "The Theory of plasticity: a survey of recent achievements Proc. Inst. Mech. Eng., 169, 42, (1955).

15. Eisenberg, M. A. and Phillips, A., "On nonlinear kinematic hardening." Acta Machanica, 5, 1, (1968).

16. Hodge, P.G., J. App. Mech., Trans. AMer. Soc. Mech. Eng. 79, 482 (1957).

17. Miastkowski J. and Szczepinski W., "An experimental study of yield surfaces of prestrained brass," Int. J. Solids Structures, 1, 189, (1965).

18. Naghdi P. M. and Rowley J. C., "An experimental study of biaxial stress-strain relations in plasticity," J. Mech. Phys. Solids, 3, 63, (1954).

19. Shield T. R. and Ziegler H., "On Prager's hardening rule," Zamp, 9, 260, (1958).

20. Bell, J. F., "Single, Temperature-dependent stress-strain law for the dynamic plastic deformation of annealed F. C. C. metals," J. App. Phys., 34, 134, (1963).

21. Rosen A. and Bodner, S. R., "The influence of strain rate and strain aging on the flow stress of commercially pur alluminum," J. Mech. Ph. Solids 15, 47, (1967).

22. Manjoine, M. J., "Influence of rate of strain and temperature on yield stresses of mild steel," Trans. ASME, 66, A211, (1944).

23. Hanser F. E., Simmons J. A. and Dorn J. E., "Strain rate effects in plastic wave propagation," University of California MRL publication, Series 133, Issue 3, (1960).

24. Marsh K.J. and Campbell J.D., "The effect of strain rate on the post yield flow of mild steel," J. Mech. Phys. Solids, 11, 49, (1963).

25. Chidister J. L. and Malvern, L. E., "Compression-impact testing of alumi num at elevated temperatures," Exp. Mech. 3. 81. (1963).

26. Lindholm, U. S', "Some experiments with the split Hopkinson bar," J. Mech. Phys. Solids, 12, 317, (1964).

27. Lindholm, U. S. and Yeakley, L. M., "Dynamic deformation of single and polycrystalline aluminum," J. Mech.Phys.Solids, 13, 41, (1965).

28. Lindholm, U. S., "Some experiments in dynamic plasticity under combined stress," Symposium on Mechnaical Behavior of Materials under Dynamic Loads, San Antonio, Texas (1967).

29. Krafft, J. M., Sullivan, A. M. and Tripper, C. F., "The effect of static and dynamic loading and temperature on the yield stress of iron and mild steel in compression," Proc. Royal Soc. Land. A 221, 114, (1954).

30. Larsen, T. L. et Als, "Plastic stress/strain-rate/temperature relations in H. C. P. Ag-Al under impact loading," J. Mech. Ph. Solids, 12, 361, (1964).

31. Malvern, L. E., "Plastic wave propagation in a bar of material exhibiting a strain rate effect," Q. App. Math. 8, 405, (1950).

32. Malvern, L. E., "The propagation of longitudinal waves of plastic deformation in a bar of material exhibiting a strain rate effect," J. App. Mech. 18, 203, (1951).

33. Lubliner, J. "A generalized theory of strain rate dependent plastic wave propagation in bars," J. Mech. Ph. Solids, 12, 59, (1964).

34. Perzyna, P., "On the thermodynamic foundations of viscoplasticity," Symposium on the mechanical behavior of materials under dymic loads, San Antonio, Texas (1967).

35. Perzyna, P. and Wojno, W., "Thermodynamics of rate sensitive plastic material," Arch. Mech. Stos. 20, (1968).

36. Perzyna, P. and Wojno, W., "Thermodynamics of a rate sensitive plastic material," Arch. Mech. Stos. 5, 500, (1968).

37. Trozera, T. A., Sherby, O. D. and Dorn, J. E., "Effect of strain rate and temperature on the plastic deformation of high purity aluminum," Trans. ASME, 49, 173, (1957).

38. Eddington J. W., "Effect of Strain rate on the Dislocation Substructure in Materials under Dynamic Loads, Ed. V. S. Lindholm, Springer-Verlag, N. Y. (1968).

39. Perzyna, P., "On Physical Foundations of Viscoplasticity," Polskiej Akademii Nauk, IBTP Report 28/1968.

40. Peters, R. W. et Als, "Preliminary for testing basic assumptions of plasticity theories," Proc. Soc. Exp. Stress Anal. 7, 27, (1949).

PART II

APPLICATION TO MECHANICAL BEHAVIOR OF METALS

1. Introduction.

In Part I of this work, a theory of viscoplasticity (of which the theory of plasticity was a part) was developed on the basis of the concept that the current state of stress is a functional of the entire <u>history</u> of deformation and temperature,

> but the history was defined with respect to a time scale which is in itself a property of the material at hand.

In particular for the case of strain-rate-independent materials under isothermal conditions, which is the topic of this part of our work, it was shown that within the restriction of small strains

$$\sigma_{ij} = \delta_{ij} \int_0^z \lambda(z-z') \frac{\partial \epsilon_{kk}}{\partial z'} \, dz' + 2 \int_0^z \mu(z-z') \frac{\partial \epsilon_{ij}}{\partial z'} \, dz' \qquad (1.1)$$

where

$$\lambda(z) = \lambda_\infty + \sum_{r=1}^n \lambda_r \, e^{-\rho_r z} \qquad (1.2)$$

$$\mu(z) = \mu_\infty + \sum_{r=1}^n \mu_r \, e^{-\alpha_r z} \qquad (1.3)$$

where λ_∞, λ_r, μ_∞, μ_r, ρ_r and α_r are positive constants and

$$z = z \, (\zeta) \; ; \frac{dz}{d\zeta} > 0, \, z \geq 0. \qquad (1.4 \text{ a, b})$$

The symbol z denotes a positive monotonically increasing time scale with respect to a time measure $d\zeta$ such that

(1.5) $$d\zeta^2 = P_{ijk\ell}\, d\epsilon_{ij}\, d\epsilon_{k\ell}$$

where $P_{ijk\ell}$ is a material tensor, which is positive definite and which, for the isotropic materials envisioned in Eq. (1.1), has the form

(1.6) $$P_{ijk\ell} = k_1\, \delta_{ij}\, \delta_{k\ell} + k_2\, \delta_{ik}\, \delta_{i\ell}$$

where k_1 and k_2 are material constants, such that $k_1 + \frac{k_2}{3} > 0$, $k_2 > 0$.

It is evident from Eq. (1.5) that $d\zeta$ is independent of the natural time scale given by a clock and thus materials described by Eq. (1.1) are strain history dependent but strain-rate independent.

The derivation of constitutive equation (1.1) was given in detail and its relation to classical theory of plasticity was examined, in some of its aspects, in Part I.

In the present paper we shall be concerned with the real behavior of metals under conditions of room temperature and slow straining. By examining data on copper and aluminum which were obtained in the laboratory by various experimenters, we shall show that Eq. (1.1) does indeed have the capability of explaining quantitatively and with remarkable accuracy such diverse phenomena as cross-hardening i.e. hardening in tension due to torsion, loading-unloading loops, and hysteresis loops during repetitive tension-unloading - compression-unloading histories, as well as behavior in tension in the presence of shear stress due to torsion.

2. Discussion of Equation (1.1).

At the present time one cannot find sufficient experimental data in the literature to determine the functions $z(\zeta)$, $f(z)$ and $\mu(z)$. Therefore in order to use the theory at all, with the object of interpreting available experimental data, we have to seek other avenues, essentially heuristic, to determine the form of the above functions.

With regard to $z(\zeta)$, we recall that the rate of dissipation γ was given by Eq. (3.15) of Part I, i.e.,

$$\theta_0 \hat{\gamma} = (\overset{\alpha}{b_1} + \frac{\overset{\alpha}{b_2}}{3}) \, \hat{q}_{ii}^{\alpha} \, \hat{q}_{ij}^{\alpha} + \overset{\alpha}{b_2} \, \hat{p}_{ij}^{\alpha} \, \hat{p}_{ij}^{\alpha} , \qquad (\alpha \text{ summed}) \tag{2.1}$$

where a roof over a quantity represents its derivative with respect to z, and p_{ij}^{a} is the deviatoric component of the tensorial internal variable q_{ij}^{a}.

For the sake of argument let

$$\theta_0 \hat{\gamma} \equiv \overset{\alpha}{b_2} \, \hat{p}_{ij}^{\alpha} \, \hat{p}_{ij}^{\alpha} \qquad (\alpha \text{ summed}) \tag{2.2}$$

Then as a result of Eq. (2.2)

$$\theta_0 \frac{d\gamma_D}{d\zeta} = \frac{\overset{\alpha}{b_2}}{(dz/d\zeta)} \frac{dp_{ij}^{\alpha}}{d\zeta} \frac{dp_{ij}^{\alpha}}{d\zeta} \qquad (\alpha \text{ summed}) \tag{2.3}$$

Equation (2.3) may now be written in the form,

$$\theta_0 \frac{d\gamma_D}{d\zeta} = \overset{\alpha}{b_2} \, f(\zeta) \frac{dp_{ij}^{\alpha}}{d\zeta} \frac{dp_{ij}^{\alpha}}{d\zeta} \qquad (\alpha \text{ summed}) \tag{2.4}$$

where, it will be recalled,

$$\zeta = \int_0^{\zeta} \{p_{ijk\ell}\, d\epsilon_{ij} d\epsilon_{k\ell}\}^{1/2} \tag{2.5}$$

i.e. ζ is a FUNCTIONAL of the strain history, and

$$dz = \frac{d\zeta}{f(\zeta)} \quad . \tag{2.6}$$

Of course if we set,

$$\theta_0 \hat{\gamma} \equiv (b_1^{\alpha} + \frac{b_2^{\alpha}}{3})\, \hat{q}_{ii}^{\alpha}\, \hat{q}_{jj}^{\alpha} \quad (\alpha \text{ summed}) \tag{2.7a}$$

then,

$$\theta_0 \frac{d\gamma_v}{d\zeta} = (b_1^{\alpha} + \frac{b_2^{\alpha}}{3})\, f(\zeta) \frac{dq_{ii}^{\alpha}}{d\zeta} \frac{dq_{jj}^{\alpha}}{d\zeta} \quad (\alpha \text{ summed}) \tag{2.7b}$$

At this point various possibilities present themselves. The simplest is to take $f(\zeta)$ = constant, in which case,

$$z = \zeta_1 \zeta + \zeta_0 \tag{2.7c}$$

where ζ_0 and ζ_1 are constants. However it can be shown (and will be demonstrated in later sections) that this choice eliminates "cross-hardening" in the sense that a change in the uniaxial stress-strain behavior due to shear prestrain cannot be accommodated. This effect has been reported. Therefore, though Eq. (2.7) is convenient, it is not very useful.

This observation applies to Ilyushin's and Rivlin's theory, where $\zeta = \zeta + \zeta_0$ and

$$\zeta = \int_0^\zeta \{d\epsilon_{ij}\ d\epsilon_{ij}\}^{1/2} \tag{2.7d}$$

Note that Eq. (2.7d) is a particular case of Eq. (2.6) ; also note the absence of the material constant k_1 and k_2 from Eq. (2.7d), which renders ζ independent of the material at hand. In order words their theories are not endochronic.

The next natural choice is to consider $f(\zeta)$ to be a linear function of ζ i.e.,

$$f(\zeta) = 1 + \beta\zeta \tag{2.8}$$

where β is a positive constant. Note that $\beta > 0$ because $b_2{}^{a} > 0$, as well as $\frac{d\gamma_D}{d\zeta} > 0$, thus necessitating that $f(\zeta) > 0$, for all ζ. As a result of Eq.'s (2.7) and (2.8),

$$z = \frac{1}{\beta}\log\,(1 + \beta\zeta) \tag{2.9}$$

an expression which has been found to give excellent agreement in the cases of some significant experiments, as will be shown in subsequent Sections.

In the absence of experimental data, the question of the form of the "relaxation" function $\lambda(z)$ and $\mu(z)$ is equally difficult.

There are two simplifying assumptions, however, which lead to a relation between $\lambda(z)$ and $\mu(z)$, so that one is left with the problem of finding the form of only one of these functions. One is that of an elastic hydrostatic response and the other is the assumption of constant Poisson's ratio.

Efficient use of the first assumption is made by writing Eq. (1.1) in terms of the hydrostatic and deviatoric components of σ_{ij}, in which case

$$\sigma_{kk} = 3\int_0^z K(z-z')\,\frac{\partial\epsilon_{kk}}{\partial z'}\,dz' \tag{2.10}$$

(2.11) $$s_{ij} = 2 \int_{o}^{z} \mu(z-z') \frac{\partial e_{ij}}{dz'} dz'$$

where K(z) is the bulk modulus. Elastic hydrostatic response implies that K(z) = = KH(z), in which case Eq. (2.10) becomes,

(2.12) $$\sigma_{kk} = 3K\epsilon_{kk}$$

The assumption of constant Poisson ratio leads to the conclusion that $\mu(z)$ and K(z) differ by a multiplicative constant, and can both be written in terms of a single functions G(z), such that

(2.13) $$K(z) = K_0 \; G(z)$$

(2.14) $$\mu(z) = \mu_0 \; G(z)$$

where G(0) = 1.

This assumption has the added advantage that, under conditions of plane stress, or uniaxial strain, the strain in the unstressed direction is related to the strains in the stressed directions by a multiplicative constant. Thus the strain increments in the direction of zero stress may be easily eliminated from the expression for $d\zeta$ so that the latter may be expressed solely in terms of the strain increments in the directions.

3. Crosshardening in Tension-torsion.

It has been observed that in aluminum and copper as well as in other metals, prestraining in torsion, well into the plastic range, has a significant hardening effect on the stress strain curve in tension.

In this Section we shall analyze data by Mair and Pugh, who have investigated this effect on annealed copper. Their experiments were performed accurately and with care, on very thin circular cylinders which were twisted well into the plastic region, so that upon unloading there remained a permanent residual shear strain. The effect of initial shear prestrain on the tensile response was then obtained by loading the cylinders in tension.

The constitutive equations pertinent to the above situation are easily found to be :

$$\sigma = \int_{o}^{z} E(z-z')\frac{de}{dz'}\, dz' \tag{3.1}$$

$$\tau = 2 \int_{o}^{z} \mu(z-z')\frac{d\eta}{dz'}\, dz' \tag{3.2}$$

where σ and ϵ are the axial stress and strain, respectively, and τ and η are the respective shear stress and tensional shear strain ; the moduli E(z) and μ(z) are interrelated through the bulk modulus K(z). Their relating is best expressed through their Laplace transforms :

$$\overline{E} = \frac{\delta\overline{\mu}}{1+\frac{\overline{\mu}}{3\overline{K}}} \tag{3.3}$$

To deal with the effect of cross-hardening analytically, we have assumed a constant poisson ratio. As a result Eq. (3.3) reduces to the form :

$$E(z) = E_o\, G(z) \tag{3.4}$$

where

$$E_0 = \frac{3\mu_0}{1+\dfrac{\mu_0}{3K_0}} \tag{3.5}$$

Regarding the form of G(z) we have taken the simplest possible view by assuming that

$$G(z) = e^{-\alpha z} \tag{3.6}$$

Despite these simplifications we have been able to obtain excellent agreement with experimental data that have hitherto lacked analytical representation.

Analysis.

In the tension-torsion test the effect of constant poisson ratio is to reduce $d\zeta^2$ to the form

$$d\zeta^2 = k_1\ d\epsilon^2 + k_2\ d\eta^2 \tag{3.7}$$

where k_1 and k_2 are material constants, not the same as those in Eq. (1.6).

During torsion ($\epsilon = 0$),

$$\zeta = k_2 \eta \tag{3.8}$$

whereas during tension ($\eta = \eta_0$) and after pretension

$$\zeta = k_2 \eta_0 + k_1 \epsilon \tag{3.9}$$

where η_0 is the maximum shear prestrain.

Equation (3.1) may now be written in the form

$$\sigma = E_0 \int_{\zeta_0}^{\zeta} G[z(\zeta)-z(\zeta')] \frac{\partial \epsilon}{\partial \zeta'} d\zeta' \tag{3.10}$$

where $\zeta_0 = k_2\eta_0$) when allowance is made of the fact that $\epsilon = 0$ in the range $0 < \zeta < k_2\eta_0$. Thus cross-hardening is taken fully into account by Eq. (3.10), through the shear prestrain parameter ζ_0, which appears as a lower limit on the integral on the right hand side of Eq. (3.10). If, in particular, we assume that $G(\zeta)$ is given by Eq. (3.6) and use of this is made in Eq. (3.10) the latter becomes

$$\sigma = E_0 e^{-\alpha z(\zeta)} \int_{\zeta_0}^{\zeta} e^{\alpha z(\zeta')} \frac{\partial \epsilon}{\partial \zeta'} d\zeta' \tag{3.11}$$

The integral in the right hand side of Eq. (3.11) can be evaluated explicitely by using Eq. (2.9) and noting that during monotonically increasing extension $\frac{d\epsilon}{d\zeta'} = k_1$. Omitting the algebra,

$$\sigma = \frac{E_0(1+\beta\zeta)}{k_1\beta n} \{1 - (\frac{1+\beta\zeta}{1+\beta\zeta_0})^{-n}\} \tag{3.12}$$

where

$$n = 1 + \frac{\alpha}{\beta} \tag{3.13}$$

and

$$\zeta_0 \leq \zeta = \zeta_0 + k_1\epsilon \tag{3.14}$$

Equation (3.12) represents a family of stress strain waves in tension, in terms of the prestrain parameter ζ_0 and the "cross-hardening" parameter (*) β.

To determine the material parameters in Eq. (3.12) we note that in the absence of shear prestrain ($\zeta_0 = 0$),

$$\sigma = \frac{E_0(1+\beta_1 \epsilon)}{\beta_1 n} \{1 - (1+\beta_1 \epsilon)^{-n}\} \tag{3.15}$$

where $\beta_1 = k_1 \beta$.

It may be verified that as $\epsilon \to 0$, $\sigma = E_0 \epsilon$ i.e. E_0 is the initial shape of the stress-strain curve. Also as ϵ increases, σ tends symptotically to the linear expression

$$\sigma = \frac{E_0}{\beta_1 n} (1+\beta_1 \epsilon) \tag{3.16}$$

(*) There is ample justification for calling β the cross-hardening parameter. Indeed in the of $\beta = 0$, and using Eq. (3.9) Eq. (3.12) becomes :

$$\sigma = \frac{E_0}{k_1 \alpha} (1 - e^{-a\epsilon})$$

which is independent of ζ_0 ; in other words cross-hardening cannot take place when $\beta = 0$, as pointed out earlier.

If E_t is the slope (tangent modulus) of the asymptotic straight line, then

$$n = \frac{E_0}{E_t} \tag{3.17}$$

Also, as shown in Fig. 1, if one extrapolates backwards the asymptotic straight line to intersect the strain axis one obtains an intercept σ_0 from which β_1 is determined by the relation

$$\beta_1 = \frac{E_t}{\sigma_0} \tag{3.18}$$

Similarly integration of Eq. (3.2) yields an equation analogous to Eq. (3.16) ; this is

$$\tau = \frac{2\mu_0(1+\beta_2\eta)}{\beta_2 n} \{ 1 - (1+\beta_2\eta)^{-n} \} \tag{3.19}$$

where $\beta_2 = k_2\beta$. Thus, β_2 and μ_0 may be determined from Eq. (3.19).

Finally we observe from Eq. (3.12) that the intercepts σ_0 in the presence of shear prestrain ζ_0 are given from the expression

$$\sigma_0 = \sigma_0 (1+\beta\zeta_0) = \sigma_0 (1+\beta_2\eta_0) \tag{3.20}$$

Equation (3.20) was used to confirm the self-consistency of the theory. However Eq.'s (3.12), (3.1) and (3.19) can only yield the ratio $(\frac{k_1}{k_2})$ but the constants k_1 and k_2 cannot be evaluated. In this sense, and for these experiments one may choose k_2 arbitrarily ; we choose $k_2 = 1$

Experimental data obtained by Mair and Pugh that illustrate the effect of cross-hardening are given in Fig. 2.

./. which is independent of ζ_0 ; in other words cross-hardening cannot take place when $\beta = 0$, as pointed out earlier.

Curve 0 is the virgin stress strain curve for the type of copper they used. The circles on the curves A, B and C are experimental points corresponding to initial shear prestrains of $.25x10^{-2}$, $1.5x10^{-2}$ and $3x10^{-2}$ respectively.

From curve 0, $E_0 = 14x10^6$ lb/ln^2, $\beta_1 = .53x10^2$, $n = 46$. With $k_2 =$ $= 1$, Eq. (3.20) was used to give $k = 1.00$. The curves A, B and C were then calculated and plotted as shown. Without a doubt the agreement between theory and experiment is remarkable.

4. Repetitive loading-unloading cycles.

The tensile strain history $\epsilon(\zeta)$ corresponding to a typical tensile loading-unloading sequence is shown in Fig. 3. We use the terms "straining" and "unstraining" in the following sense :

The ranges $0 < \zeta < \zeta_1$ $\zeta_2 < \zeta < \zeta_3$, $\zeta_B < \zeta < \zeta_5$ $\zeta_D < \zeta$ represent straining in tension.

The ranges $\zeta_1 < \zeta < \zeta_2$, $\zeta_3 < \zeta < \zeta_A$, $\zeta_5 < \zeta < \zeta_C$ represent unstraining in tension.

The ranges $\zeta_A < \zeta < \zeta_4$, $\zeta_0 < \zeta < \zeta_6$, $\zeta_7 < \zeta < \zeta_8$ represent straining in compression.

The ranges $\zeta_4 < \zeta < \zeta_B$, $\zeta_6 < \zeta < \zeta_7$, $\zeta_8 < \zeta < \zeta_D$ represent unstraining in compression.

Points on the ζ-axis denoted by ζ_r (r-1,2....) represent points of discontinuity in $\frac{d\epsilon}{d\zeta}$, brought about by reverting from straining to unstraining histories, or vice-versa.

A persual of experimental data on copper, shows that the constitutive equation of the metal varies depending on its previous history of manufacture and subsequent annealing. The single term form of $G(\zeta)$ that explained Mair and

Pugh's data (2) so well was found inadequate to explain data by Lubahn (3) and by Wadsworth (4).

We found however, that the adoption of a single extra term for G(z) sufficies to describe quantitatively broad trends of their data. In effect we took

$$G(z) = G_1 + G_2 e^{-az} \tag{4.1}$$

or

$$E(z) = E_1 + E_2 e^{-az} \tag{4.2}$$

Let ζ_m (m = 1, 2....) be the last point of discontinuity in $\frac{d\epsilon}{d\zeta}$. Then using Eq.'s (2.9), (3.11) and (4.1) and in the range $\zeta_m < \zeta$

$$\sigma = (1+\beta\zeta)\sigma_0 \left\{(-1)^m - \frac{1}{(1+\beta\zeta)^n} + 2 \sum_{r=1}^{m} (-1)^{r+1} \frac{(1+\beta\zeta_r)^n}{1+\beta\zeta}\right\} + E_1\epsilon \tag{4.3}$$

The quantities ζ_r may be evaluated explicitely in terms of ϵ_r (the values os strain corresponding to ζ_r) by the formula

$$\zeta_r = 2k_1 \sum_{s=1}^{r} (-1)^{s-1} \epsilon_s + k_1(-1)^r \epsilon_r \tag{4.4}$$

The effect of E_1 on the unstraining characteristics is remarkable, especially since its effect on the shape of the straining part of the stress-strain curve is minimal. Let the history $\epsilon(z)$ be one of continuous straining. Then Eq. (4.3) becomes :

$$\sigma = E_1 \epsilon + \frac{E_2(1+\beta_1\epsilon)}{n\beta_1} \{1-(1+\beta_1\epsilon)^{-n}\} \tag{4.5}$$

From Eq. (4.5) we obtain the following relations in the notation of Section 3.

(4.6a) $$E_1 + E_2 = E_0$$

(4.6b) $$\frac{E_2}{\beta_1 n} = \sigma_0$$

(4.6c) $$E_1 + \sigma_0\beta_1 = E_t$$

Equations (4.6a-c) do not suffice for the determination of the four unknown material constants E_1, E_2, n, β_1. It has been found that a fourth realtion can be obtained by considering the "unloading" portion of the stress strain history.

Fig. 4, shows the stress-strain relation for a uniaxial specimen which has been strained in tension to a strain value ϵ_1 ; whereupon it is unloaded and compressed until the final strain is zero.

The strain-intrinsic time measure history $\epsilon(\zeta)$ corresponding to the above stress-strain history is also shown in Fig. 5.

Equation (4.3) in conjunction with the above history yields the relation, at $\epsilon = 0$:

(4.7) $$\sigma = \sigma_0(1+\beta\zeta)\ \{ -1+\frac{2(1+\beta\zeta_1)^n-1}{(1+\beta\zeta)^n} \}$$

If the value of ϵ_1 is sufficiently large (in the case of copper this value was found to be $50x10^{-3}$, or so) then $\sigma_0{}^c$ is given very nearly by the expression

(4.8) $$\sigma_0{}^c = -\sigma_0(1+2\beta\zeta_1) = -\sigma_0(1+2\beta_1\epsilon_1)$$

The constant β_1 can now be obtained from Eq. (4.8) and the constants E_1, E_2 and n can be found from Eq.'s (4.6a-c).

We illustrate the points made in the above discussion in Fig. 6 where stress-strain curves for three different materials are given when these are subjected to the same strain history shown in Fig. 5.

The constants for these materials are given in the following table :

	E_1	E_2	β	n
1	0	6.14×10^6	$.4 \times 10^2$	25
2	$.24 \times 10^6$	5.9×10^6	0	∞
3	$.12 \times 10^6$	6.02×10^6	$.2 \times 10^2$	50

What is remarkable is that changing β results in these materials having indistinguishable stress-strain curves during straining but wildly differing ones during unstraining.

In Fig. 7 we illustrate an attempt to predict analytically the loading-unloading-loading response of copper in simple tension. The solid line is an experimental curve obtained by Lubahn (3) for a copper specimen which had already undergone similar strain cycles. We have assumed, however, that these have a negligible effect in the response shown because they occured sufficiently far in the distant "past".

The traingular points shown, were obtained theoretically from Eq. (4.3) by assuming that the specimen was continually extended (without unstraining) until the strain $\epsilon = 51.6 \times 10^{-3}$ was reached. The unstraining-straining cycle was then applied.

Despite the fact that E(z) was approximated by two terms, as in Eq. (4.2) the agreement between theory and experiment is remarkable. The constants employed were, $\sigma_0 = 6 \times 10^3$, $E_1 = .125 \times 10^6$, $\beta = .02 \times 10^3$, $n = 160$.

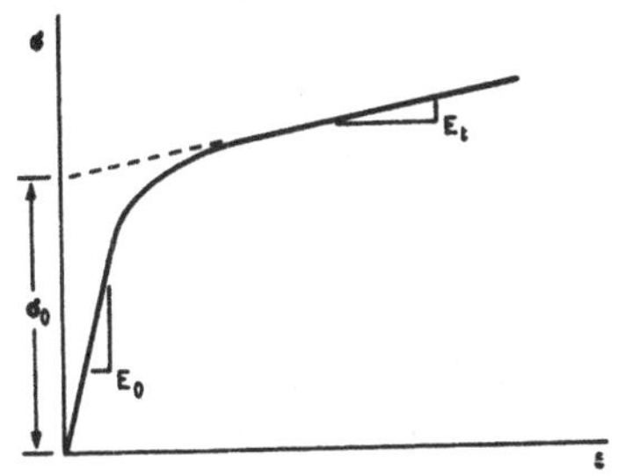

Fig. 1. Typical stress-strain curve.

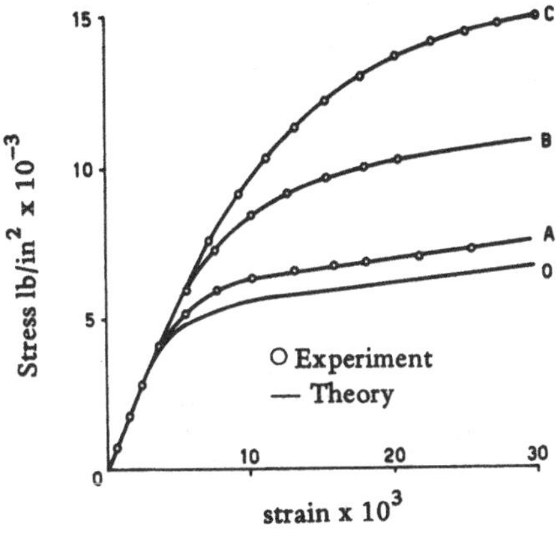

Fig. 2. Hardening due to shear prestrain

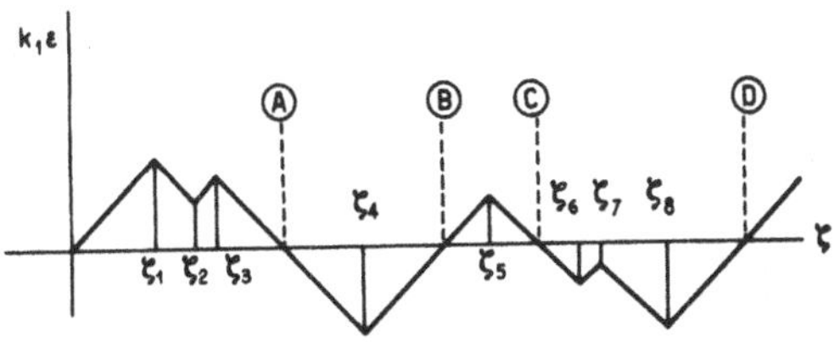

Fig. 3. Typical loading - unloading - loading sequence

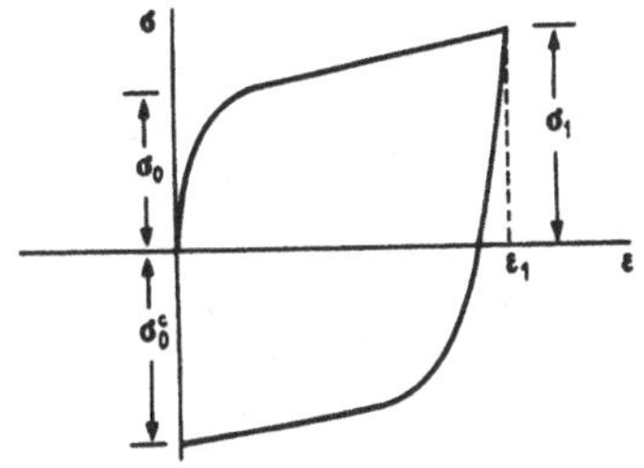

Fig. 4. Tension - Compression test

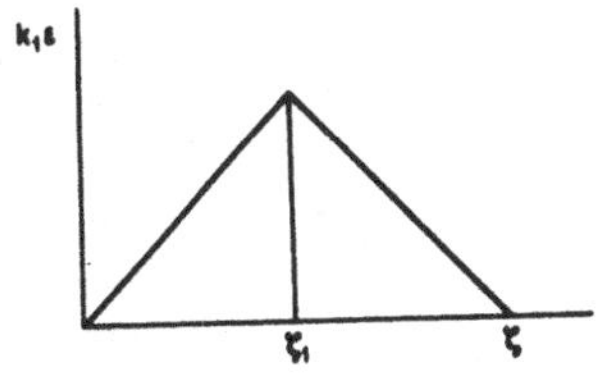

Fig. 5. Strain history of Fig. 6.

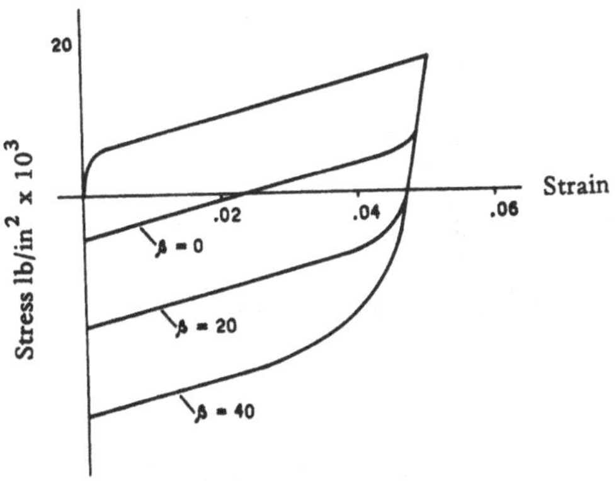

Fig. 6. Bauschinger effect and its relation to β

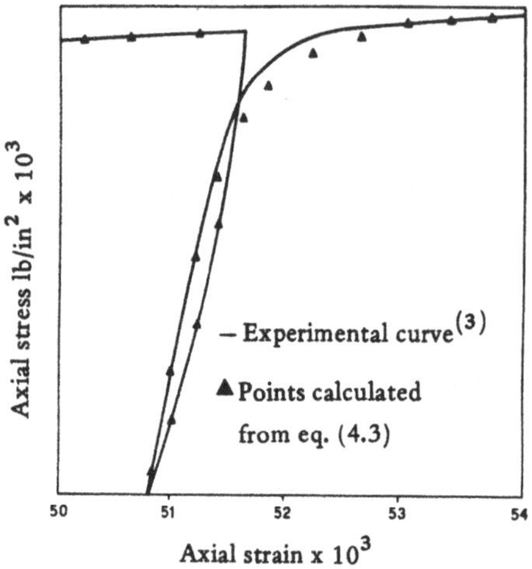

Fig. 7. Unloading - loading loop

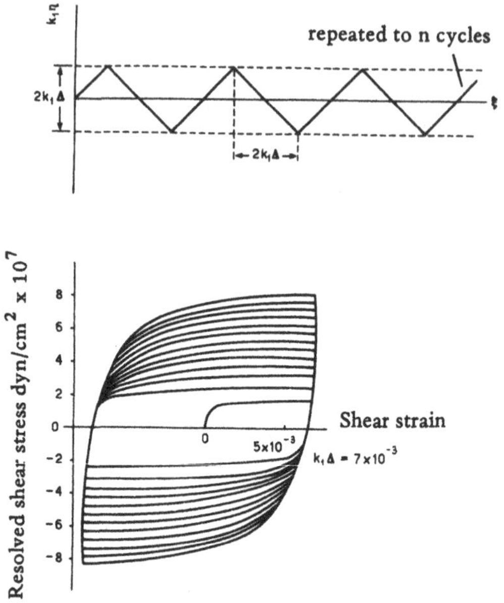

Fig. 8. Cyclic straining

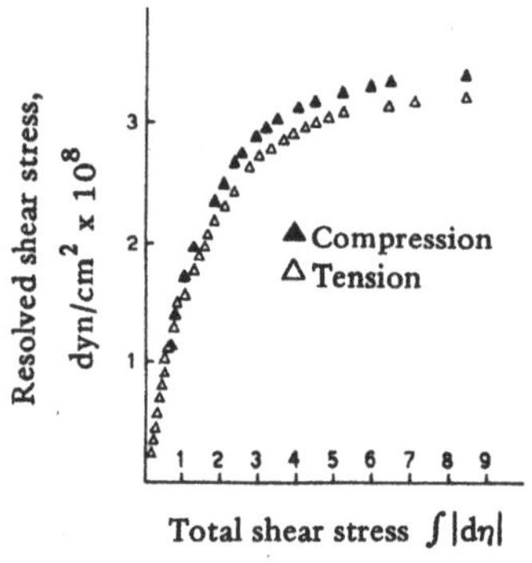

Fig. 9. Peak stresses due to cyclic straining

Fig. 10. History of resolved shear strain

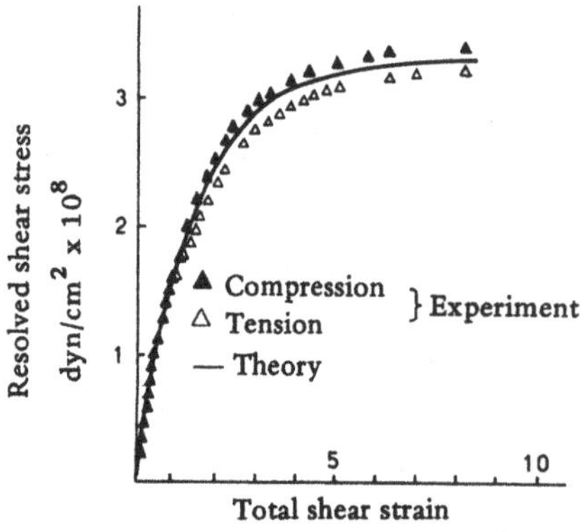

Fig. 11. Theoretical prediction of cyclic hardening

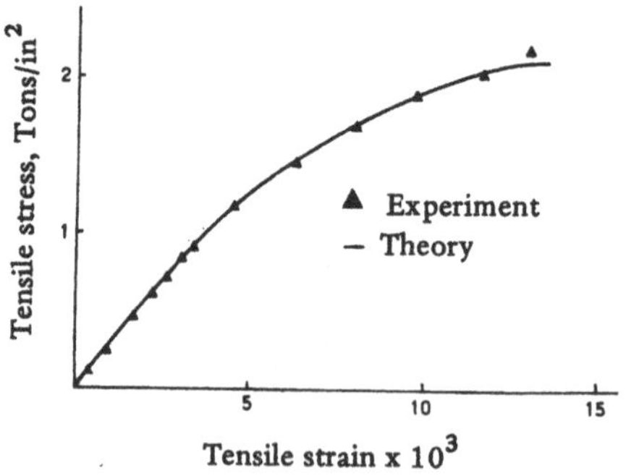

Fig. 12. Theoretical prediction of effect of shear prestress

In fact we are not aware of another instance where an attempt was made to describe such experimental data analytically by means of one single constitutive equation. In addition we can say with assurance that the observed difference between theory and observation can be reduced further by including more exponential terms in the series representation for E(z). We conclude this Section by considering the effect of work hardening under cyclic straining. In particular we shall examine the work of Wadsworth's (4) and show that our theory again provides an excellent analytical basis for his results.

In this work single copper crystals were tested under conditions of uniaxial cyclic strain. The data was presented in terms of the resolved shear stress and strain in the plane of slip.

Fig. 8 gives the first few cycles of his straining program, in which a crystal was cycled under fixed limits of resolved shear strain of 7×10^{-3}. The "peak stresses" corresponding to the extreme values of tensile and compressive strain increased monotonically with the number of cycles.

In Fig. 9 the values of peak tensile and compressive stresses have been plotted by Wadsworth against $|d\eta|$. It is rather interesting that he felt that such a plot was meaningful, without further elaboration on this point. Of course $|d\eta|$, but for a scalar factor, is our intrinsic time measure.

The history of the resolved shear strain versus ξ is shown in Fig. 10. From this Figure it follows that $\xi_m = (2m - 1)\Delta k_1$. Equations (4.6c) and (4.8) were now utilized to find βk_1, which we denote by β_1, and E_1. It was found that $\beta_1 = 12.3$ and $E_1 = 2 \times 10^9 \frac{\text{dyn}}{\text{cm}^2}$. At this point n could not be determined because the initial slope of the stress-strain curve corresponding to $\xi < \xi_1$ could not be evaluated accurately.

However letting $\tau_m \equiv \tau|_{\xi = \xi_m}$, it was found that as $m \to \infty$, Eq. (4.3) yields the asymptotic expression :

$$\tau = n\beta\tau_0\ \Delta + E_1\Delta \tag{4.9}$$

Hence, from the tensile experimental curve of Fig. 4 of Ref. 4, n could be determined explicitly and was found to be equal to 225. For this value of n the term $\frac{1}{(1+\beta\zeta_m)}$ n was found to be negligible for m>1.

Thus for the history in Fig. 10, Eq. (4.3) gives

$$\tau_{m+1} = \tau_0[1+\beta_1(2m+1)\Delta]\,(-1)^m + 2\sum_{r=1}^{m}\left[\frac{1+(2r-1)\beta_1\Delta}{1+(2m+1)\beta_1\Delta}\right]^n (-1)^{r+1} + E_1\epsilon \tag{4.10}$$

The above equation can be simplified further for large values of m. In particular for m ≥ 50, it was found that the series in the bracket on the right hand side of Eq. (4.10) degenerates into the geometric series

$$\sum_{r=1}^{m}(-u)^r \sim -\frac{1}{1+u} \tag{4.11}$$

where

$$u = \left\{\frac{1+(2m-1)\beta_1\Delta}{1+(2m+1)\beta_1\Delta}\right\}^n \tag{4.12}$$

Equation (4.10) may now be written in an asymptotic form in terms of the absolute value of the shear stress as follows :

$$|\tau_m| = \{\ 1+\beta(2m+1)\Delta\ \}\,\tau_0\ \{-1+\frac{2}{1+u}\} + E_1\Delta \tag{4.13}$$

For every large values of m (m >> n) Eq. (4.13) simplifies further and becomes

(4.14) $$|\tau_m| = \frac{n\beta_1 \tau_0 \Delta}{1-\frac{n}{m}} + E_1 \Delta$$

Thus,

(4.15) $$\text{Lim}|\tau_m| = n\beta_1 \tau_0 \Delta + E_1 \Delta$$

In Fig. 11 a plot has been made of the theoretical relation between $|\tau_m|$ and $m\Delta$ obtained from Eq. (4.13). The experimental points obtained by Wadsworth are also shown. The following comments are in order. Though our theory does give values for τ_m which are different in tension from those in compression, the difference is not as great as the experimental data indicate, and is too small to be plotted on the scale shown. However, the theoretical curve lies very close to, and is in fact bounded by the experimental points, which indicate a deviation between the values of compressive stress and those of tensile stress which increases with m but is never greater than 5.5%.

This is the first time that a theory of plasticity has provided a rational explanation for the phenomena of cyclic hardening.

5. Tensile response in the presence of initial shear stress.

In Section 3 we obtained a theoretical prediction of the effect of pre-strain in torsion on the stress strain curve in tension. In this Section we shall examine theoretically, in the light of our endochronic theory, the effect of initial constant prestress in torsion on the stress-strain curve in tension. To do this, we have assumed, just as we did in Section 3, that E(z) and μ(z) are proportional to some relaxation function G(z), and furthermore that G(z) consists of a single exponential term i.e. it is given by Eq. (3.6). Thus

$$E(z) = E_0 e^{-az} \tag{5.1}$$

In the light of Eq. (5.1) and bearing in mind Eq. (2.9), the constitutive Eq.'s (3.1) and (3.2) can be reduced to the differential equations

$$E_0 \frac{d\epsilon}{d\zeta} = \frac{a\sigma}{1+\beta\zeta} + \frac{d\sigma}{d\zeta} \tag{5.2}$$

$$2\mu_0 \frac{d\eta}{d\zeta} = \frac{a\tau}{1+\beta\zeta} + \frac{d\tau}{d\zeta} \tag{5.3}$$

where, as in Section 3,

$$d\zeta^2 = k_1 d\epsilon^2 + k_2 d\eta^2 \tag{5.4}$$

As mentioned above the test to be discussed consists of applying an initial stress τ^0 corresponding to an initial strain η_0 ; then keeping τ^0 constant, a axial strain ϵ is applied and the axial stress σ is measured. The object at hand is to deduce from Eq.'s (5.1 - 5.4) the relation between σ and ϵ, and compare with the experimental data obtained by Ivey (5).

To accomplish this we proceed as follows. From Eq. (5.4) it is clear that the exial straining process begins at $\zeta = \zeta_0$ where

$$\zeta_0 = k_2 \eta_0 \tag{5.5}$$

During this process $\frac{d\tau}{d\zeta} = 0$, so that from Eq. (5.3)

$$d\eta = \frac{a\tau^0}{2\mu_0} \frac{d\zeta}{1+\beta\zeta} \tag{5.6}$$

Equations (5.4) and (5.6) now combine to show that during the axial straining process,

$$d\zeta^2 = k_1^{\ 2}\, d\epsilon^2 + k_2^{\ 2} \left(\frac{a\tau^0}{2\mu_0}\right)^2 \frac{d\zeta^2}{(1+\beta\zeta)^2} \tag{5.7}$$

At this point we introduce the variable θ such that

$$\zeta = k_1 \theta \tag{5.8}$$

Also let $\frac{k_2}{k_1} = k$, $k_1 \beta = \beta$ and $c = (ka\tau^0/2\mu_0)$. Then, in terms of θ and as a result of Eq. (5.7)

$$d\theta \left\{1 - \frac{c^2}{(1+\beta_1\theta)^2}\right\}^{1/2} = d\epsilon \tag{5.8}$$

Equation (5.8) may be integrated subject to the initial condition that at $\zeta = k_2\mu_2$, $\epsilon = 0$; or, $\theta = \theta_0 = k\eta_0$, $\epsilon = 0$.

Equation (5.2) may now be integrated with respect to θ to yield

$$\sigma = \frac{E_0}{(1+\beta_1\theta)(a/\beta_1)} \int_{\theta_0}^{\theta} (1+\beta_1\theta')^{(\alpha/\beta_1 - 1)} \sqrt{(1+\beta_1\theta')^2 - c^2}\, d\theta' \tag{5.9}$$

We introduce now a change of variable by the relation

$$1 + \beta_1\theta = c \cosh\phi \tag{5.10}$$

whereupon Eq. (5.9) becomes :

$$\sigma = \frac{E_0 c}{\beta_1 \cosh^{n-1}\phi} \int_{\phi_0}^{\phi} (\cosh^n\phi' - \cosh^{n-2}\phi')\, d\phi' \tag{5.11}$$

where as before, $n = 1 + \frac{a}{\beta_1}$.

Now,

$$\int \cosh^n n \, dx = \frac{n-1}{n} \int \cosh^{n-2} x \, dx + \frac{1}{n} \cosh^{n-1} x \sin x \qquad (5.12)$$

Since for asymptotically larbe n(say $n > 30$), $\frac{n-1}{n} \sim 1$, it follows from Eq. (5.12) that in this instance

$$\int (\cosh^n x - \cosh^{n-2}) \, dx \sim \frac{1}{n} \cosh^{n-1} n \sinh x \qquad (5.13)$$

The result of Eq. (5.13) can be utilized to obtain a closed form solution for σ which now becomes,

$$\sigma = \frac{E_0 c}{\beta_1 n} \sinh\phi - \left(\frac{\cosh\phi_0}{\cosh\phi}\right)^{n-1} \sinh\phi_0 \qquad (5.14)$$

Equation (5.8) may also be integrated with respect to ϕ to yield

$$\epsilon = \frac{c}{\beta_1} \{F(\phi) = F(\phi_0)\} \qquad (5.15)$$

where

$$F(\phi) \equiv \sinh\phi - \tan^{-1} (\sinh\phi) \qquad (5.16)$$

Thus σ and ϵ are related parametrically through ϕ and ϕ_0 such that

$$\phi_0 = \cosh^{-1} \left\{\frac{(1+\beta_1 k\eta_0)}{c}\right\} \qquad (5.17)$$

The relation between σ and ϵ has been calculated with the following values of the constants :

$$k = 1, \ \sigma_0 = \frac{E_0}{\beta_1 n} = 17.4\text{x}10^3 \ \text{lb/ln}^2, \ \beta = 20, \ n = 33.$$

The result was compared with one of Ivey's experiments in which $\tau_0 = 14 \mathrm{x} 10^3$ $\mathrm{lb/m}^2$, $\eta_0 = 2.35 \mathrm{x} 10^{-3}$. The predicted and experimental stress-strain responses compare very favorably. See Fig. 12.

Conclusion.

On the evidence of the results presented above it appears that the endochronic theory of plasticity can predict accurately the mechanical response of metals under complex straining histories. The full implications of the theory will be investigated further in our future look.

REFERENCES

1. Valanis, K. C., "A theory of Viscoplasticity without a yield surface, Part I-General Theory", Mechanics Report 1.01, University of Iowa, Dec. 1970.

2. Mair, W. M. and Pugh, H. Ll. D., "Effect of Prestrain on Yield Surfaces in Copper", J. Mech. Eng. Sc., 6, 150, (1964).

3. Lubahn, J. D., "Bauschinger Effect in Creep and Tensile Tests on Copper", J. of Metals, 205, 1031, (1955).

4. Wadsworth, N. J., "Work Hardening of Copper Crystals under Cyclic Straining", Acta Metallurgica, 11, 663, (1963).

5. Ivey, H. J. "Plastic Stress Strain Relations and Yield Surfaces for Aluminum Alloys", J. Mech. Eng. Sc., 3, 15, (1961).

CONTENTS

Part II : Application to mechanical behavior of metals

Printed in the United States
By Bookmasters